AF730999

OLIVES NATURALLY

WITH THE PUGGLE EXPRESS

By

SIMON FIELD

Farm Fandango

The fandango is a lively and passionate Spanish courting dance. The dancers challenge, tease and pursue each other, a reflection of the courtship of wildlife and the ingenuity of the evolved lures used to attract a mate or pollinator. As custodians of the farm, we observe the plants and creatures, their intimate interaction, the ebb and flow of the interrelationships, and the constant changes in rhythm with the four seasons. We are participating in a perpetual fandango.

In the words of St Francis of Assisi, 'O Lord of love and kindness, who created the beautiful earth and all the creatures walking and flying in it, so that they may proclaim your glory, I thank you to my dying day that you have placed me among them'. The 'puggle express' is a delightful annual mating rite of the Australian marsupial echidna which is used here to demonstrate the benefit of developing a farm as an ecologically balanced enterprise. This has been achieved through establishment of an organic olive grove and associated activities which provide adequate income and a platform to test and promote a different, less invasive, approach to farming.

The base is the physical farm comprising remnant vegetation, wetlands, a 6,500 tree olive grove, pastures for grazing sheep, and human habitation. On this foundation has been built a table olive processing enterprise, an international business in training and testing olive oil profilers, consulting in all things olive oil and table olives, and publishing.

The income from these off-farm activities allows the optimisation of the production from the farm without pushing it to its limit using artificial inputs. The story that follows relates the intimate relationship this approach has created between the fauna, flora, and humans.

About the author

Simon Field was born and spent his early years in Zimbabwe. He studied Natural Sciences at Trinity College, Cambridge University, and holds a Master of Arts. After graduation he travelled the world on his own before returning to the farm. For ten years he farmed, married, and raised a family.

Looking for new opportunities, the family migrated to Australia in 1981 and started a farming enterprise which they still own. Following some involvement at the senior representative levels of farming organisations he was appointed the Executive Director of the Australian Institute of Agricultural Science, a position held for 8 years. In 1997 he resigned from the Institute and planted a 5,000 tree organic olive grove on the family farm.

He has written handbooks on almost every aspect of olive oil and table olive production and marketing. He has presented numerous papers at conferences worldwide and written many articles on the production, marketing, and consumption of olive oil and table olives. In 2023 he published the books 'Fruity, a Little Bitter and Spicy – A producer's view of the global olive oil industry' and 'African Olive', a discourse on the southern African olive industry. In 2024 he published 'The Savantes Complete Olive Handbook – Olive oil and table olives production to sales'. His family and other animals continue to farm olives in St Arnaud, Victoria, in south-eastern Australia.

Cover art by Benjamin Field,

Four years old and student at St Michael's Montessori Nursery School, Elm Bank Gardens, Barnes, London SW13 0NX

Written by Simon Field

Compiled by Michael Field

1362 Dunolly Road

PO Box 104
St Arnaud
Victoria 3478
Australia

Email: simon@savantes.org

ISBN. 978-0-6457327-5-7

For Annie, Joss, Michael, and Roger who helped change this small part of the earth, and the animals and plants which allow us to share their world

and

Benji who is a future custodian of our precious planet.

Table of Contents

Preface

Looking out of the window as I write on an overcast spring morning there are kangaroos, or 'mowers' as we call them, grazing on the grass in front of the house with a backdrop of rolling pasture dotted with tall grey box trees. When we started farming here in 1981 the outlook was very different with unbroken views of bare eroded paddocks as far as the eye could see to the horizon of blue hazy hills, and little or no grass for the mowers.

The transformation is not confined to our property, there has been a massive shift of thinking in the custodianship of the land by farmers, starting locally with the effort of a friend, Terry Simpson, who started Landcare with a community project of revegetation on the severely eroded Stricta Hill at Winjallock in the St Arnaud area of Victoria, in south-eastern Australia. Subsequently hijacked by politicians as their initiative with little continuing recognition given to the grass roots volunteer initiators, Landcare grew to a national movement which has been pivotal in the reconstitution of ecosystems. I accept that the involvement of politicians and associated bureaucracy is necessary to elevate a local initiative to a national movement, however, I question the usurping of the idea for political gain and status with the true initiator being marginalised. The earth bears greater witness to Terry Simpson's ecological initiative than the accolades absorbed by the passing parade of politicians.

Bureaucracy has taken the place of weekend community tree planting parties at Stricta Hill where townspeople and farmers joined to take collective responsibility for the environment. Taxes are now collected and redistributed for Landcare projects, replacing the volunteer labour and provision of plates of food and eskies of beer as a reward for the toil. As is so often the case, government has institutionalised land restitution and removed the nexus between the local inhabitants and their surroundings.

Gone are the challenges between the owners of various utes as to which would prevail in a pushing competition in an arena made of applauding locals. At Stricta Hill I think the Landcruiser won and no damage was done!

There is a cost to changing stewardship of farming land from short term profit driven monoculture to a longer term optimal integrated agriculture which may yield less annual return but delivers a more self-sufficient farming business. If one interprets the word profit to mean 'gain', it opens up many more income streams, not just those emanating from delivering a truckload of grain to the silos or a trailer of wool to the auctions. The opportunities include off-farm activities built on the knowledge or services garnered from farming, vertical integration to access more income from downstream processing of product, and valuing the intangibles such as lifestyle and ecosystems.

Coming from an interest as a child in entomology, I followed the scientific path through school and university, and ended up with a degree in Natural Science giving a good foundation to understand the interaction of the fauna and flora on the farm. Purchasing the property as recent migrants in 1981 we were obliged to establish from scratch a farming enterprise and opted for what we knew by building an intensive pig production business. For various reasons, not least of which was the unrelenting labour and thirst of the enterprise for increasing capital investment for uncertain returns, we closed the piggery down after 10 years. There was relief from no longer subjecting the animals to a life on concrete confined in crates, keeping them healthy with feed infused with antibiotics.

Off-farm activities included involvement in farming organisations at a representative level in the state body and national committees. It was a very active period in farming politics with the farmers flexing their muscles in numerous stoushes with unions which were inflating costs of manufacturing, shearing, and the use of ports. The activities unified the farmers and victories increased their self-esteem and earned greater respect from both political allies and opponents. The National Farmers' Federation Fighting Fund was generated to stand as a warning to militant unions that the farmers' union had the financial capacity and the will to take them on in the courts and elsewhere. It has been used to great effect over the last 40 years.

We kept the farm and I embarked on a second career in our new country as the Executive Director of the professional body for agricultural scientists, The Australian Institute of Agricultural Science. Within a year I also became the editor of the professional magazine Agricultural Science and the delegate for Australasia to the Commonwealth Association of Agricultural Scientists, 'CASAS'. These positions exposed me to the full range of opinions on all matters agricultural, including the tension between sustainability and productivity. The membership of the Institute of Agricultural Science represented every aspect of the industry – academics of all disciplines, bureaucrats, environmentalists, consultants, researchers, educators, farmers, and business people covering commercial supplies of agricultural inputs and sales of all products.

At the beginning of my time with the organisation I regarded the representations of the nascent environmental movements as suspicious and anti-farming. As we started to develop official policies to promulgate the interests of the input suppliers of herbicides, pesticides, inorganic fertilisers, and all manner of artificial inputs and controls, the environmentalists became more fraught. As an institute I believed it was incumbent on us to foster reasoned debate on important issues. An example was genetic engineering of herbicide resistant cereals and cloning of animals. It was the time of the outbreak of mad cow disease and we convened a seminar entitled 'Mad Cows and Mutants – Are They Worth the Risk?' It was boycotted by the National Farmers' Federation as being too sensationalist and it incurred the wrath of the agricultural chemical industry which began moves to ensure their interests were well represented in the Institute hierarchy. I saw this as positive in that more open and objective debate would take place.

Around the same time, as the delegate from Australasia, I attended a conference in Singapore convened by CASAS. I was already uncomfortable with the unbridled promotion of proprietary chemicals associated with genetically engineered cereals and the purchase of crop seed banks by multi-nationals to consolidate control by some of the world's largest agricultural chemical conglomerates. At the conference I attended a presentation by a scientist from one of these multi-nationals on the herbicide glyphosate and the patenting of 'ready' modifications of soybeans and other crops which were resistant to it. At this stage I didn't understand the full implications of this development but I was very disturbed at the exhortation by an employee of the conglomerate developing these patented seeds that they would need the support of scientists to get them accepted. The implication was that if agricultural scientists supported the new products they would be rewarded with research funds and other incentives. The overarching argument was that the increased crop yields would help green and feed the world more efficiently – the undercurrent was the users of the seeds would be locked into contracts with the suppliers thereby giving unprecedented control over global cropping.

That presentation confirmed my misgivings about the unrelenting push to maximising production using an ever increasing array of patented inputs which will potentially upset whatever ecological equilibrium still existed in agricultural practices. The financial implications also bothered me as contracts would tip the balance firmly in favour of input suppliers. Added to this, the long term implications of this disruption of the natural systems had not been adequately modelled, partially because there were so many variables and unknowns. I resolved to find another way.

The course to this was not through the Institute where success in policy promulgation had increased relevance of the organisation, and interests which were threatened sought to assert more control over policy direction. I resigned, partly for this reason and because I have this biorhythm with a seven year cycle of change. Unsurprisingly my position was taken by a former employee of the largest agricultural chemical company in Australia, in a way back-handed recognition of my success in opening up the debate to a more environmentally considerate approach.

It was back to the farm to test some nascent theories. The following pages chart the progress of the farm as an ecologically balanced enterprise with an associated 'let nature find its own equilibrium' attitude. This approach tests one's resolve when there are so many quick fixes offered for every agricultural event which is perceived as a problem.

The descriptions are fashioned in plain English rather than technical jargon requiring each species to be taxonomically identified. We don't know the scientific names of many of the species and families referred to, and we are sure our readers don't either, knowing the common name. We have a library of books which identify birds, flowers, trees, grass, butterflies, and other insects. Dipping into these to present a reasonable façade of knowledge of the species quickly brought us to the conclusion that it is too hard. There are thousands and often the differences are the domain of taxonomists, botanists and entomologists, not mere earthlings!

Our observations may seem naive to some ecologists and scientists, hopefully in some cases they might learn from our interpretations which have an element of emotion common to relationships between creatures. We could search pages of books and websites to arm ourselves with the specific detail of the flora and fauna, while it may be edifying, we prefer to be outside observing and being amused and delighted by the creatures who act on impulse and their social norms without needing to identify, qualify, and quantify everything they see and do.

The Puggle Express

In mid-September as the gloomy overcast days of winter are melted by growing warmth of the onset of spring, an amorous event happens. We call it the 'Puggle Express' or the 'Love Train'.

Its advent is heralded by a constant swishing noise under the house as a female echidna takes up her post in a cosy corner and awaits the arrival of the first of a train of suitors. This is followed by the 'hot' pursuit which culminates when the male with the greatest stamina, and perhaps guile, claims his prize as they join in procreation. The males can be seen converging from all directions guided by an amorous scent which dulls the mind to any distraction from impending conjugal bliss. The traditional door to the open love nest is the gap under the step at the back entrance of the house. Stepping out I have to be careful not to tread on the spiny suitors who are all consumed with anticipation of participating in the 'Train' and perhaps being rewarded with a sideline shunt!

When enough males have gathered the female leads them off on a merry journey which can cover hundreds of metres and take a few days. The order of march for the followers changes as the older, stronger ones jostle the inexperienced youngsters out of the way and to the back of the queue. Finally the novices and unfit tire and drop off and only one or two of the most persistent remain, whereupon she takes her pick and sneaks off with him to consummate the love ritual.

The exercise is so Darwinian with the survival of the fittest that the evolutionary winner is the perpetuation of this enchanting small monotreme for an estimated 20-50 million years. The only other living monotreme, or egg-laying mammal, is the platypus which lives in an aquatic environment. Thought to have evolved from a platypus-like antecedent, echidnas are also good swimmers.

The annual 'love-in' gathering under the house is testament to the lack of fear for the human inhabitants of the farm. We have even been privy to the final copulation on numerous occasions, something about which the animals are very coy and is the subject of talk-back radio discussion where various theories are postulated – like the female digging a trench and lying on her back to submit in the 'missionary' position. The ground here is too hard for that and there seems to be an anthropomorphic element to this suggestion. Our observation is that the participating parties lie on their sides and the lengthy male penis is more than adequate in bridging the spiny divide. They lie in this position for a while, perhaps exhausted by the chase which may take days and involve up to 10 suitors, or just enjoying the sexual experience enough to keep them fulfilled for the year between love trains. Or perhaps a bit of both. We confess to being a little embarrassed by our voyeuristic recording of their naked intimacy with the cameras on our smartphones.

Once fertilised the female produces an egg and a 'puggle' hatches into a pouch where it stays for a while before growing spines and being evicted. The mother digs a burrow where the puggle is deposited and returns to feed the growing offspring for up to a year. Then the puggle emerges and is on its own to find a territory to patrol for food.

Our farm is around 220 hectares and the longest puggle express we have seen is seven. Often when we are walking around the property we see an echidna out feeding or resting away from the heat in one of their many lairs in old tree stumps. Individuals range quite widely, regularly covering their territory and visiting favourite feeding areas. Devouring copious amounts of ants, their visits are marked with disrupted ant's nests. A sighting of one of the resident monotremes gives us a lift and enhances our day.

A recent news item reported that after observing echidnas for many years scientists had recorded them communicating with grunts and wheezes. Prior to this we had filmed and recorded a lone echidna digging and emitting a series of grunts and other noises which could be characterised as communication. As the echidna was fully aware of my presence I like to think it was communicating with me as a fellow inhabitant of his territory!

Another recent discovery is that to keep cool from the searing Australian sun, echidnas blow snot bubbles and perform belly flops on cool surfaces. The snot bubbles burst and wet an area on the tip of the nose where blood accumulates just under the skin to be cooled by evaporation. The belly flopping seems to be a back-up for the snot routine, presumably because cool surfaces are hard to find in the bush on a hot summer day. Recently we observed this cooling activity in a fryng pan placed at ground level and filled daily to provide water for low slung creatures like lizards, goannas, echidnas, and even snakes. Late spring and summer had been unusually dry with the gound parched, dams drying up and food for all very scarce. In the late evening on an extremely hot and humid day I looked out of the window from my cool air-conditioned room and noticed an echidna ambling towards the bird bath. Assessing that the water was too high to reach it headed off to the full frying pan and after a long drink of water climbed in. It did some belly dancing and flicked water over its back with its feet. After a lot of splashing it settled low in the pan and wallowed for a good twenty minutes. Slowly it eased its way back onto the ground, rubbed its belly in the sand, and then ambled off in the direction of the olive grove and a night of ant consumption. This was a rare and delightful episode to observe and elicited a silent commitment to always fill the frying pan on a hot day. Unbearable is the thought of our spiny resident walking all the way to the house to find the lazy inhabitants hadn't filled the pan on a day with the sun blazing down relentlessly.

Recognising their ability to dig rapidly to expose some ants or anchor themselves in the soil with only the spines exposed to predators, we have called the largest male Kubota and the females the Ditch Witches. Giving the inhabitants of the farm a quirky name helps us identify them, albeit anthropomorphic.

This year the puggle express attracted only one male in the vicinity of the house. Perhaps an unexplained reduction in the ant population has forced the local residents to search further afield for their food, or the female has decided to advertise her desires elsewhere. We have seen one echidna as we emerge from their winter hibernation, the sighting coming with relief that the population has not been decimated.

Pollinators, Predators and Pests

There is a variety of ants for the echidnas to vary their meals. The largest are the bull ants. Big and black, these can be up to an inch long and deliver a very painful sting. They protect their small nests aggressively and it seems even the spiny ant eaters give them a wide berth. Walking barefoot across the grass our grandson received a bull ant sting and a deep respect for ants is now etched into his mind for life, he can spot an ant's nest at 20 metres. Meat ants inhabit low-lying nests marked by a patina of grains of sand. Step on one of these and suddenly the soldiers emerge in their hordes from numerous holes and attack. Sometimes when working and concentrating on the olive trees we inadvertently stand on a nest and suffer many bites on the lower body. In the case of a serious attack a rapid and unashamed disrobing occurs as the ascending perpetrators are located and eliminated, each releasing an acrid odour of formic acid as they are squashed.

There is a somewhat bizarre connection between dead ants and olives. Social insects such as bees, ants and termites have a mechanism for removing their dead to minimise the risk of infections or parasites gaining a foothold in their nests or hives and causing mass death. Oleic acid, resulting from the interruption of metabolic pathways in the dead insects, acts as a trigger for a hygiene response from the living whereby the bodies are disposed of in ant middens or tossed out of hives. Some insect communities have a more rapid response than others giving them an advantage in surviving entomological pandemics. Oleic acid is the predominant monounsaturated fatty acid found in olive fruit to which is attributed medicinal benefits, especially in the mitigation of cardiovascular disease in humans. In the one case the chemical compound is a call for the undertakers, in the other it is claimed to be a life saver. Some have described the odour of dead ants as reminiscent of olive oil, one presumes rancid oil rather than the fresh aromas of extra virgin olive oil. The mind boggles at the science which may be needed to untangle the chain reaction when ants consume other insects which have feasted on olives and are high in oleic acid!

Many species of small black ants infest various areas and we are assailed by workers when harvesting or pruning the olive trees. Feeding on the honeydew generated by the sap-sucking scale insects on the trees the ants form a constant stream up and down the tree trunks from the food source to their earthly nests. At times there is so much honeydew on the leaves that the trees glisten as if glazed with sugar. This makes it very hard to harvest the olives as they stick to gloves and will not drop into the buckets. Wetting the gloves works but is not an attractive prospect in the winter harvest when temperatures are regularly in single figures.

The ants actively farm the scale to ensure the supply of honeydew by killing other insects such as ladybirds, aphytis wasps, and spiders which predate on the scale. Who said humans were the first to farm?

Two years ago, an extraordinary event occurred – the ant population in the grove almost disappeared. Coincidentally the scale insect infestation diminished having been a constant nuisance for many years. We are uncertain about which came first, the death of the scale depriving the ants of honeydew, or the death of the ants removing the protection from the scale which were attacked by other predators. We can only assume that some scale or ant pandemic occurred, and the insects were wiped out over 12 months. It makes working the trees easier, but some consequences may increase the amount of pruning required to control other insects.

There are many predators of scale including certain wasps, hover flies and ladybirds. With the omnipresence of scale we saw a huge increase in the ladybird population and then when the scale virtually disappeared, so did the ladybirds.

Early on in the establishment of the grove as an organic ecosystem we detected the arrival of the scale insects from a small number of trees on the western edge. From there the infestation steadily spread to all parts of the grove and seemed unstoppable. Starting in 1997 we experienced a 12-year drought with the olive trees becoming severely stressed from the lack of water. Despite having a large dam which we used for supplementary irrigation, the lack of rainfall resulted in all the five dams on the farm drying to a floor of cracked clay. It was so dry we carted in water for our homes and the houses shifted on their foundations as the soil was depleted of the elasticity provided by moisture. Our reference to the drought on the internet resulted in the inclusion of pictures of the desiccated farm in a school book in Jamaica as an illustration of the effects of the long term dry. All the necessary permissions were sought and granted and we hope the students find the inclusion edifying.

The olive trees in their extraordinary resilient way curled their leaves to shut down water loss from transpiration. The drop in sap deprived the scale of their source of nutrition and the insects died. When the rain came back so did the scale as well as new natural predators called aphytis wasps. The wasp eggs are laid in the scales and devour the victim during development. Other predators such as lacewings appeared and eventually a large population of ladybirds decorated the trees. Now, twenty-five years after the establishment of the grove there are few scale insects left. If we dared think that the natural evolution of the grove ecosystem had brought the insect life to equilibrium, we were wrong!

We have had three very wet years, the wettest for the 40 years on the farm. The retreat of the scale and the humid weather has brought in another insect plague. A sap-sucking insect, the olive lace bug, has gratefully taken the place of the scale on the underside of the leaves and sucked the sap giving the leaf a yellow spotted surface. Suicidal in its proliferation the bug can infest all the leaves on the tree and result in severe leaf drop. We first noticed this terminal tendency in a few small, deserted groves close to the farm where all the trees appeared to be dying and losing leaves. Like so many leaf discolorations on our trees we noticed the yellowing effect and put it down to any one of; dieback, trace mineral deficiency, waterlogging, and many others. Then they really attacked one tree and started to develop 'hot spots' as the infestation spread from tree to tree. Now we are busy hacking off the affected branches to prevent the next cycle of infestation which starts in spring with the eggs that have overwintered producing the first of a potential four generations in summer. The problem has been so challenging that a chapter has been devoted to our efforts to control the lace bug. How much easier it would be to spray insecticide to solve these insect attack problems and how devastating this would be to the wonderful ecosystem we have encouraged on the farm which will be described in the chapters of this book.

This is an example of the fine balance of the insect population where we have not used pesticides, fungicides, herbicides, or bactericides for over 25 years. The interdependence of the fauna and flora is constantly evident with the seasons and weather variations of dry and wet adding another dimension. We are yet to see if the decline in the ant population has an effect on the number of echidnas which gather each year to join the 'Puggle Express'. Last year there were only two suitors, well down on previous years, and this spring there was no love train. As stated earlier we are unsure whether this decline is caused by a reduction in numbers or other love trains taking place at other locations around the farm. We hope it is the latter, even if it means the return of the blitzing by ants when we work on the olive trees.

Termites are the silent destroyers of wood. If it happens to be the wooden stumps of your house or structural timber this is a bad thing. On the other hand their activity returns wood to the earth as a natural fertiliser and humus. Many Australian hardwoods are used for fence posts as they are so hard that even termites turn to softer ligneous vegetation. Generally we have not experienced any detrimental effects of termite activity in the olive grove. Having heard of their ability to destroy cashew nut trees by devouring the dead wooden centre of the trunk we are always on the lookout for the telltale 'soil tunnels' which mark their presence. We have found them working their way up dead bark and removed the protective tunnels to expose them to the birds which were happy to devour them. Within a very short time the termites disappeared and hopefully will stay away – touch wood! There are plenty of them around the farm working on fallen logs and branches returning them to the earth in an almost funereal way, like a slow natural cremation without the heat.

The bird population on the farm has exploded with the increased vegetation and copious amounts of food. Here we will mention a few instances of the interaction between birds and insects while dealing with the bird population and their antics in an expansive way later. There are no behavioural silos amongst the fauna, one cannot describe one species without recognising the effect it has on the other inhabitants of the ecosystem.

Back to insects. The temptation is to adopt the conventional agricultural term 'pests', which is unfair. They are no more pests to us as humans are to them. Pest seems to be the collective term used for anything that is animate and a nuisance to our cosseted lives where we try to control everything to suit our purpose. In the same way weeds cover anything that is growing where we don't want it to grow. Paradoxically, most of the vegetables which we farm were weeds which we tamed and modified to our taste with sales pitches which emphasise their medicinal properties and digestive benefits. Meanwhile, animals graze and browse selectively to get these benefits directly without the need for an extended supply-chain, shelves of expensive supplement pills, and associated promotional campaigns.

Another weed which we cultivated to give us pleasure was tobacco, now seen as an evil killer of millions of humans every year. On the neighbouring farm which has sandy soil there is the ruin of a tobacco barn, evidence that this weed was once grown here for commercial gain. We haven't come across plants of the other weed used by some for recreational pleasure or pain mitigation, namely cannabis. We have regular grid searches overhead by police helicopters presumably looking for illicit crops hidden in the forests which make up the surrounding patchwork of national parks.

I won't refer to the curculio beetle, or apple weevil, as a pest, just a challenge when it devours large quantities of fresh olive leaves. This canny little creature lives in the soil at the base of the tree during the day and scuttles up the trunk when the birds and other predators are asleep at night and has a feast. Evidence of its meal is the sawtooth like serrations around the edge of the leaves, as if it has been trimmed with pinking shears. A few of these nocturnal raiders is tolerable but like all things they don't know when to stop and multiply uncontrolled until they denude a tree of all its vegetation.

Casting around for a way to stop these nightly forays we heard of an orchardist who had tied a wad of fibrous material, such as raw wool, around the trunk. The weevils got caught up in the fibres on their journey up or down and subsequently expired. We tried this with wool and it was very successful, until the troop of black raven-like birds called choughs found that there was a free breakfast every morning on the wool girdled trees. They proceeded to dismantle the girdles to find any embedded weevils and the surrounds looked like a repository for wool waste. While we pondered how to overcome this new development we observed that the wise choughs had worked out that the beetles spent the day in the soil. They now dig around the base of the trees when the weevils are abundant and we no longer have a problem with trees being denuded overnight! The ability of birds such as choughs, ibis, and magpies to hear the movement of prey in the soil is impressive. A foraging magpie will step slowly across the grass, stop, listen intently then plunge its beak into the soil with great accuracy and extract a long wriggling earthworm which it devours. Likewise, a platoon of ibis with their long, curved beaks will move in formation across the paddock plunging them into the ground to find some tasty morsel.

There is always noise around the farm, and in the olive grove in particular. The farm has an area which has been fenced off for over 30 years and is only accessible to native wildlife. Registered under the Land for Wildlife scheme, the fauna and flora are untouched and there is a substantial area of forest which has been left wild for much longer, unmolested apart from some trees being felled by previous owners for fence posts.

The olive grove with approximately 6,000 trees covers about a quarter of the remaining land. The balance is open grazing land interspersed with some dry creeks which are now lined with native eucalyptus trees. When we acquired the property in 1980 the creeks were bare of vegetation and eroding. Our laissez-faire approach to natural revegetation has resulted in the creek lines being well vegetated providing habitat and corridors for all manner of creatures. Sometimes the noise made by a flock of cockatoos and corellas in these riverine tree lines is so loud that it is difficult to think.

One seasonal sound is the buzzing of bees harvesting the pollen from the olive flowers in late spring. They busily load their legs with pollen and fly off to wherever the hive is to deliver their cargo. These are the honey bees and many will come from commercial hives which are dotted around the area which is well known for 'yellow box' honey. Yellow box is a native tree of the eucalyptus family. Paradoxically the olive trees are wind-pollinated and the bees do not have an important role in pollination. They do make a lot of noise while going about their work – so much that we notice when some years they are notably absent. There may be many reasons for this – a preferred source of ingredients for honey, a decline in the population, hives being relocated, and I am sure there are others. It is disquieting when there is a Rachel Carson 'silent spring'. We wonder what has happened, knowing that it is not our application of pesticide which has decimated the insects.

On a hot summer's day, in unison with the shimmering heat waves, male cicadas exercise their tymbal organ to produce a high pitched sound to attract females, protect territory, warn off predators, or just talk to each other. Like the traffic noise for people living near freeways, we seem to be able to tone this background noise down to tolerable levels.

We do not live in isolation and the activities of neighbours can affect our semi-contained ecosystem. Birds fly in and out carrying a myriad of seeds and eggs and the wind will blow in a constant flow of flying fauna. The neighbours who agist their sheep on the grazing land bring in hay and grain to supplement the grass for sheep. All of these activities can have an impact on our wildlife. The olive grove is quarantined to all except us – in this way we keep some control over the across boundary traffic. We have no tractor or tillage equipment, working out ways to do everything by hand with as little impact on the environment as possible.

There are native bees which are present and essential for the pollination of native flora. A recent visit by a city-based son located five hives of native bees high up in cavities in various eucalyptus trees. I had never noticed them in my many years on the farm – probably because I had not looked for them. Showing greater interest after the discovery we learn that there are many species of native bees carrying out all manner of essential work in the ecosystem. The bees will be added to the list of fauna protected on the farm.

Moving silently throughout the warmer months are the butterflies and moths. Being a budding entomologist in my younger days I have to express disappointment at the lack of variety of butterflies compared to my previous home in the African sub-tropics. There we had many species which exhibited every colour combination of a painter's palette and included swallowtails and fast flying charaxes, with a myriad of smaller yellow, white, blue and multicoloured combinations. On the farm we have the meadow argus, painted ladies, cabbage and caper whites and many little blue butterflies. On a few occasions something more interesting has winged its way to a flowering gum for a passing visit, but that is about it!

The moths are no more interesting, mainly nocturnal they escape our scrutiny so there are probably more than we know. We are alerted to the movements of the bogong moths which, attracted to the house lights, commit suicide by thumping into the windows. Breeding in regions around our location the bogongs have an annual pilgrimage to aestivate – or enter a dormant state – for the summer in the cool of mountain caves in the Victorian Alps to our east. Archaeologists found evidence in Cloggs Cave of the use of the moths 2,000 years ago as a food source for the Aboriginal population. The proliferation of habitation and the accompanying lighting along the route the moths take to and from the mountains seems to have disrupted the annual migration. The endangered mountain pygmy possum, which can be found around Australia's highest mountain, Mount Kosciusko, feeds on the bogong moth and is consequently being deprived of one of its food sources. This has prompted some conservationists to advocate moderated use of lighting during the critical months to create dark corridors to the mountain caves.

Every day there is a new discovery or an amusing cameo presented by the wildlife. New colourful vistas are painted by the seasonal change of the many layers of flora – the yellow of the wattle, the pink of the fringed heath myrtle and the white daphne heath – all against the backdrop of the blue-green eucalypts and rolling pastures of winter green or summer brown grass.

Lace Bugs Test Our Organic Resolve

Every year brings different seasonal combinations which test our resilience in managing the olive grove without resorting to the quick fix of artificial inputs. Most of the time we get through with a reasonable yield and a manageable workload. The seasons spanning 2023 to 2025 have proved by far the most challenging.

The start of the season back in November 2023 was auspicious with good rains accommodating bounteous flowering and promising olive fruit set. The trees which produced a lot of fruit the previous year started to grow in all directions and the trees which had an off year set a lot of fruit. We are happy to have the heavily laden trees as long as the irrigation and summer rain replenish the soil water reserves. The trouble with the rapid growth of many trees is they become too big for us to handle without machinery, and they steal the water from the smaller trees with the fruit as the root growth matches the above ground foliage growth.

The olive trees are not the only things which enjoy a humid and cooler summer, the insects thrive in the conditions. We noticed adjacent groves showing symptoms of water-logging and dieback, so much that the entire groves looked dead. A few of our trees started showing the same dieback, and on closer inspection we identified the cause as massive infestations of the olive lace bug. Research told us that the insects are uniquely Australian and have migrated from the native mock olive *(Notelaea longifolia)* to the cultivated olive. The mock olive is in the same taxonomic family as the European olive with similar looking fruit. It grows along the hot and humid fringes of rainforests covering the coastal region of northern New South Wales and Queensland. Its liking for these conditions may explain why it has thrived in the humid heat of this season. No doubt its spread across all regions will result in adaptation to drier summer conditions and it will become a threat to productivity in more normal summers with hot days with low humidity.

Having the bug as our own means that the research which would be done by the main olive growing countries around the Mediterranean is limited and we must rely on local research. There is even less research for organic control methods other than cutting off the branches bearing leaves which the insect has infested, or using 'organic' pesticides which will also eliminate other insects.

Subsequent research revealed that similar bugs infest olive groves in the Western Cape region of South Africa, and the plane lace bug is attacking the plane trees lining the streets of London in the United Kingdom.

In the last two seasons we have learned a lot, the first being that controlling the spread of the bug in a 6,000 tree olive grove takes a lot of planning and hard work. The bug infests the tree by progressing from a clutch of eggs laid near the tip of the underside of the leaf. The hatchlings go through various stages to the winged mobile adult. The adults use whatever means they can to travel to other leaves and trees to lay their eggs. They will fly assisted by the wind, walk along branches, attach themselves to kangaroos brushing the tree, hitch a ride on birds which alight in the tree and humans who are pruning. The research suggests they cannot fly great distances, but somehow they have successfully infested every single tree or olive grove we have seen this season. Visiting a friend's winery, a quick inspection of their few olive trees showed well progressed infestation. The bugs are everywhere.

While they can appear on trees not adjacent to infested trees, generally inspection of adjacent trees shows they spread from the 'hotspot' like ripples on a pond. The trick is to closely inspect the trees and prune off any branches which have eggs or later developments of up to five nymphal instars. Adults can be seen singly on leaves and these are also removed.

The breeding life cycle can be 3-7 weeks and varies on different olive varieties. There will be some synchronisation of the infestation in spring when the eggs, nymphs and adults which have successfully over-wintered restart the life cycle. Thereafter through summer there can be all stages of the life cycle throughout the grove. There is evidence that cold winters no longer break the cycle as effectively as in the past, as some have survived to reinfest early as the weather warms up. This is a not so gentle reminder that we do not deal with static ecological situations as the fauna and flora adapt to changing conditions.

Learning all this, our management has progressed from dealing with the bug in a haphazard way to trying to go through the entire grove in a three week period, cutting off infected branches on the periphery of the tree, and chain sawing all the branches of severely infected trees. Coincidentally this treatment has removed eggs and bugs which we subsequently learned take refuge in the bark to survive the winter.

The grove now has buffer areas around the original infested trees where all the branches have been removed from adjacent trees. Unlike the tactics used by our governments to control the spread of covid, we cannot isolate the trees and impose a curfew on the birds and kangaroos! Creating buffer zones is the next best option and seems to be helping. The removal of foliage also eliminates the shading effect so the kangaroos which lie in the shade during the summer keep away from these areas and eliminate themselves as a vector, the same applies to the birds.

Over three months it appeared we had brought the bug infestations back to the eggs and nymphal instars and reduced the progression to adults. We have been helped by the natural predators like the ladybird, green lacewings, soldier beetles and an array of spiders, including the golden orb where we find adult bugs in their web middens. Some recent hot dry weather has helped to slow down the bug breeding, while unfortunately causing the olives on unirrigated trees to shrivel. Our inclination is to accept the shrivelling if the dry helps control the bugs.

Scale insects are also helping control the bug by competing for space on the leaves where the eggs are laid and exuding the very sticky mildew which with condensation on the humid nights forms a sticky syrup which coats the leaves, especially the leaf tips where the bugs congregate. The sticky kills the nymphs and stops the adults from flying, like superglue. The trees are so sticky that we have to wash our pruning secateurs and gloves daily. In normal years, if there are such things, there would be a myriad of ants feeding on the mildew and protecting the scale from predators such as the ladybird. This season there are virtually no ants so the mildew flows freely. The ants would also eat the olive lace bugs.

Coming to the end of the second summer we hoped that our drastic measures to control the bug would allow us to reduce the infestation to controllable levels. During winter after harvest we have pruned all the trees with special attention to any evidence of lace bug eggs. Now at the end of summer we have found very few infestations. Like the disappearance of the ants the bugs have gone back to normal numbers. We find leaves with evidence of sap sucking but no sap suckers. It appears we have the bug under control without using any pesticides. Despite a long, dry, and hot summer the trees look really healthy and the reduced foliage from the heavy pruning has limited the impact of low soil moisture. In spring we will diligently inspect every tree to ensure that the wily insect does not catch us napping for another year.

To quote from the book 'An African Love Story' by Dame Daphne Sheldreck, founder of the David Sheldreck Wildlife Trust, "He (David) taught me to trust my instinct and common sense rather than seek complicated explanations to Nature's riddles. He said 'You can always count on one thing and that is Nature usually knows best and can provide the best solution under any set of circumstances, for Nature is adaptable. Don't ever be fooled into believing that man must be obliged to play God, because when he does, he usually messes things up and triggers another set of problems"'.

Researching the lace bug and paying such close attention to each tree has taught us more about the interaction and interdependence between the trees and insects, birds, marsupials, and arachnids. We have learned to recognise the eggs and breeding cycles of soldier beetles, green lacewings, scale, and ladybirds – all predators and our allies. We have frequently discussed the quick fix of spraying pesticide – hitch the fan sprayer to the tractor, fill the tank with pesticide, and in two hours clean up the bug. And clean up everything else, upsetting the ecological balance we have striven to establish. Instead we have spent many hours a day over two seasons trying to contain the outbreak. The result of the discussion is to keep at it, have confidence in the natural way and keep learning. Our yields will be down and our income will be reduced. Our minds will be at peace as we continue to enjoy the olive grove and all the creatures we share it with!

When we drive past the groves which were infested at the same time as ours and see the total devastation caused by the third year of infestation, we reflect on the hours spent pruning the bugs off and allow ourselves a moment of self-congratulation. Only a moment as Nature has a way of asserting itself over those who see it as anything other than an ally in achieving mutually acceptable results.

Songs and Feathers

The noise of the birds provides a wall of sound, almost orchestral in its diversity, and the volume would impress Phil Spector. Any video taken of an animal will have on playback a background noise of birds communicating, a constant melodious symphony of many instruments. There are songs and cries of the different species and then the different calls for each species – alarm, territorial, mating, conversational, mimicking, practising, and many others known only to themselves. The carolling chatter of the ubiquitous magpies in the olive grove in the morning reminds me of the animated greetings and exchange of news of those emerging from the New York subway after an overnight stay. Oblivious to the surrounds and the hardship of the night, they welcome the new day emerging into the fresh air and impending new experiences. They garrulously recount the events of the passed night to their fellow subway dwellers, matters trivial to some but immensely important to them in their socially barren lives.

The magpie is regularly voted Australia's most popular song bird. This year is an exception when the willie wagtail came fifth outscoring the magpies which came seventh. We can hear wagtails chiding their black and white competitors through the night reminding them that they are on their way down the slippery poll!

The wagtails in the garden constantly harass the eight magpies which regularly fly in for food and water. Sitting on the paddock fence, one wagtail attacks the birds as they glide in from the adjacent copse, while its mate lies in wait in the garden trees then attacks the magpies on the ground. Then they work together giving their targets little time to relax as these irritating and nimble little black fantails attack from behind. Occasionally the irritation levels result in the the magpie retaliating, by which time the wagtail is long gone, perched in the tree mocking the irate bird for its lack of agility.

Hearing the magpie warble is like coming home. They are aggressive in the breeding season swooping on passers-by, sometimes inflicting damage and often terrifying children on the way to school. Their reputation for terrorising populations wins headlines in national and international news with graphic pictures of screaming children scuttling down the street chased by a swooping bird. It is said that they are adept at facial recognition and are less severe on friendly faces and never forget an adversarial face. We know their nesting spots on the farm and anticipate the swoops, delighting in the news that a new brood is on its way.

A year ago we were driving back from Melbourne in the late afternoon when a magpie swooped on the car on Fogarty's Gap road. It dived right in front of the car hitting it with a thump. I looked in the rear view mirror to see if it lay damaged on the road or had escaped unscathed. There was nothing to be seen, so we drove the next hour back to farm. When I parked the car I noticed that the front grill had been broken and behold there was a magpie claw and some feathers protruding – and they moved. Very carefully I extracted the extremely indignant magpie which had a very unpleasant ride for 100km at 100km per hour in the front seat. Fogarty, as it was to be called, squawked and semi-flew, apparently undamaged. Recaptured, it spent the night in a covered cage to recuperate before being transported back to the site where the car was attacked. When released it flew off with an angry squawk to no doubt tell all its mates the story. Some superglue repaired the car grill and we look for Fogarty when we frequently drive along that road. To release it on the farm would have meant almost certain death at the beaks of the resident magpies and there may have been some orphaned chicks on Fogarty's Gap road.

Various individual magpies have thrust themselves into our lives. The first was Thatcher, appropriately named after the Iron Lady. We regularly get strong squally winds in spring and the magpie chicks get blown out of their minimally constructed nests. The adults may use anything they can find, especially if it is shiny, to fabricate their nests, however they wouldn't get any awards as builders. Their nests are little more than a rickety platform of twigs high up in the trees.

Thatcher was found on the ground, wet, wingless and in a bad way. Pocketing him or her – we never found out which so we will assume her – the near lifeless chick was taken to the warm house to dry out and be tended. The first magpie orphan taught us a lot about what to do and what not to do. First there was research and calls to local animal carers about the diet which was administered with a nasal dropper. Slowly solids were introduced and the young bird started to exhibit a character of its own and grow sturdier. The wing feathers grew then broke. They dropped out and were replaced with a stronger set and tentative fluttering started. Living in a cage overnight and mainly on top of it during the day, Thatcher became the friend of our two young boys and the dog. She was most indignant when the elder son built a magpie from Lego and put it adjacent to the cage. This was an invasion of space and not popular. It did not cower when intimidated and this added to the indignation of the territorial aggressor.

Every morning an outside session was introduced and invariably, in being carried from the top of the cage to the front door, there would be a very liquid bowel movement which landed on the furniture or carpet. No matter how quickly we made the transfer she managed to time the morning defecation to perfection!

Learning to run she played in the garden and made it difficult to catch her at nightfall – until she had enough and then demanded to be collected by standing in the flower bed and tapping on the window with her beak. Then she refused to be put into her cage until the clock struck 10pm! Slowly she took some tentative flutters as her wing feathers and muscles grew stronger and finally one day she took off accidentally and, squawking as only a magpie can, did a circuit of the house, tried to land in tree, missed, and spiralled down to the ground in a flurry of feathers. Seemingly embarrassed she cleaned her beak and preened as she contemplated her latest adventure. Her flying progressed until she cleared the fence and was gone into the nearby trees. From then she returned to the wild, coming in for the odd meal before she became just another member of the local magpie mob. Sad to lose this intimate experience with a bird with so much character, we delighted in the return to her natural environment and grateful for the insight she had given us into the behaviour of a great songbird. By playing with the boys, she instilled a lifelong love in them for the species and this has extended to all animals.

Thatcher was the first of three orphans reared in the home. The next was Parkinson who was similarly blown from the nest in a storm, and the third was Peggy who had broken legs from an inadvertent stomp from the dog's paw. The local vet repaired the legs with splints and curled claws were opened up with a thong-like homemade shoe. Peggy walked again, but with a limp, and flew just fine, eventually joining the others in the magpie families around the farm.

Frequently one hears a cacophony of sound comprising the alarm calls of magpies, cockatoos, wattle birds, and others as they mob the grandest of our bird population – the wedgetail eagle. Apparently the fourth largest eagle species in the world these majestic creatures appear on the farm frequently and soar effortlessly high above, riding the thermals and covering great distances. They have a particular tree which they favour as a lookout for prey and one day we were graced with the sight of the adults and three offspring devouring a kangaroo carcase. They are beautiful birds with menacing piercing eyes and an air of majesty. Covering a substantial area as their territory we do not know where they nest but wait to welcome the new brood when they bring them to the hunting ground. With the increase in fauna on the farm the visits are becoming more frequent and it is our hope that one day they will nest in the forest at the back of the property. We have named the male Zeus and the female Rhea after the mythical Grecian gods.

There are other raptors around the farm, kites, falcons, and sometimes little eagles. One day on a walk around the forested area I heard frenzied high pitched screeching and carefully moved towards the source. It came from a small raptor dismembering its prey on the thick trunk of a fallen tree. As it tore strips of flesh it screamed in delight, or perhaps showing its hunting prowess to others. A few days later when the feeding frenzy was over I inspected the tree and there was a bloodied area resembling a butcher's block where this hunter regularly brought its prey for the last rites.

When we had a visit from the local wildlife and conservation officer we were walking in the bush area when a falcon flew at speed screeching and changing direction in a trice. Commenting that I wondered what it was up to, the officer said it was probably flying for the thrill of flying, a concept which had not entered my mind. Since then I have learned more about the behaviour of birds and understand that they have the same range of behavioural patterns as we do, hunting, feeding, learning, playing, competing, fighting, with an added one which we don't have unless in an aircraft, flying for pleasure.

The next in the avian pecking order are the parrots, the largest of which are the cockatoos and corellas. The latter with the yellow plumes roam in flocks and feed on newly planted cereal seeds or dig out the onion grass in the paddocks. They have immensely strong beaks and crack open the very hard olive pits to extract the kernel. To amuse themselves they strip the bark off the olive trees and sit high on the television aerials snapping or bending the alloy components so the aerial looks like Boris Johnson's hair. Needless to say the TV reception is either cut-off or severely affected.

The stripping of the bark on the eucalyptus trees where they nest in cavities has a purpose, it makes the tree trunk too slippery for goannas and other lizards which scale the trees to steal the eggs or young chicks. Some trees are ring barked and die making them excellent nesting prospects for all sizes of parrots. One such tree has such a high occupancy rate that we call it the Dakota after the chichi New York apartment block. We will drive past observing cockatoos, galahs, rosellas, and red-rump parrots standing sentinel outside their 'apartments' on various 'floors' on the Dakota, with a pair of galahs occupying the penthouse. The Dakota is an example our ability to preserve habitat. Ringbarked after years of attention from various parrots the very large box tree died and, being close to the power lines, needed to be removed lest branches fell across and caused a fire. The local power line maintenance company sent in their contractor, appropriately named 'Skyrider', with his equipment to lop the tree. Rather than fell the tree, after some discussion, he was persuaded to lop the dangerous main branches and leave the trunk intact. Rising high off the ground the tree is now the seasonal home for many species of parrots.

The cockatoo flocks often contain many slightly smaller corellas. The latter are also white, have no sulphur plume and seem to have make-up around the eyes. They are no less adept with their beaks and together these large parrots sit high up in the trees like Christmas decorations and make a raucous tuneless noise. One can be quietly going about one's business when a cockatoo flies over and lets out an alarm call which is loud enough to give a heart attack.

One weekend there was a different noise from some cockatoos in front of the house. Being interested in all things that happen, which are unusual, I investigated and found that five cockatoos had cornered a ringtail possum at the base of one of the very large grey box trees and were attacking its tail, which was bleeding. The birds took it in turn to peck at the defenceless possum stopping it from escaping up the tree. Disturbed by my intervention the birds took off and the possum scrambled up the tree taking refuge in a fork. Left alone, the cockatoos returned and resumed their attack, presumably seeing the possum as a threat to their nests. Hearing the squawks again, I took a compost bin with a hatch down to the tree and placed it at the base. Poss the possum scurried down the tree and headed straight into the safe haven. How it knew to do that I do not know. The bin became its refuge for some days and we could lift the lid off to check on it and see that's its tail was healing. We got a possum box made by 'Happy', the local conservation officer, and nailed it high up in the tree so Poss could make it a permanent home. It hasn't occupied the mansion yet and we no longer see it around which we consider to be good.

The galahs are a large parrot with a pink chest, grey body, and pink plume. They pair for life and gather in flocks or traverse the skies in their pairs. If an adult is killed by a car or other catastrophe, its mate will be seen grieving and trying to revive it. Apparently they do not seek another partner and remain single for life. The grieving is proof that birds do have feelings for each other and can express emotion.

The galahs consume nuts and grains and are often seen intermingled with the flocks of cockatoos and corellas. They gather together in formation and fly with incredible precision at great speed, the sun catching the pink of their chests and then switching to grey as they wheel as one, often low overhead, and there is a whoosh as they pass. When a low pressure system is imminent heralding rain, the galahs get up to all sorts of antics, hanging upside down on power lines or tree branches, performing acrobatics with a characteristic screech, and turning somersaults as they strut across the ground. On a dead tree in front of the house we have been able to observe their courting behaviour without interference, watching them sidle up to their mate and ‘kiss’ and cuddle much as humans do. Sometimes we observe a tiff or a challenge for mateship, all confirming the emotional relationships within this species. Some inner-city people often refer to country people, especially farmers, as ‘galahs’ in a derogatory patronising way. To me this is a badge of honour as the collective behaviour of galahs, their civility, and their spirit for life is far more attractive than some of the self-interested, avaricious behaviour of some of the inner-city elites.

We had one rescued baby galah which was in a poor way with a wound on its wing and unable to fly. It screeched so loudly when handled it was named Decibel. The ungrateful fledgeling would show its displeasure at being handled by biting a finger with its vice-like beak, often drawing blood. Cosseted indoors overnight with its wound dressed with saline, Decibel spent the day in an outdoor aviary specially built for recuperating avian patients. All the time the galah was in captivity its parents sat in the trees around the aviary and reassuringly talked to it. Eventually Decibel recovered and walked out of the aviary and with a few tentative flaps left the ground, cleared the fence, and joined its caring parents in the tree.

Handling the larger parrots takes some skill as the beak which cracks nuts and olive pits could easily sever a finger. There is much advice from wildlife carers and on the internet on diets and looking after damaged or displaced animals. The alm must always be to nurse the patient back to good health and return it to the wild in the location where it was found. Those are the surroundings with which it is familiar and the parents are never far away, most often taking their young back under their wing, so to speak.

There are many smaller parrots which come in a beautiful array of colours. The rosellas in particular are strikingly decorated. The eastern rosella is a Picassoesque canvas of red, yellow, and green, and the crimson rosella is a blend of crimson, blue and green. The purple capped parakeets screech with delight as they raid the flowers of the lemon scented gum, busily going about their work like a cardinal with a purple 'zucchetto'. Occasionally we are visited by the rainbow bee-eater which nests in tunnels in the banks of the creek, a humble abode for such a strikingly beautiful bird. The most common of the smaller parrots are the budgerigar-like red rump parrots which scour the ground in groups picking up the grass seeds and grains which are their fare.

The large dead trees dotted around the farm provide homes for these parrots which can be seen perched in pairs along the branches. An eastern rosella, which we unimaginatively called Rosy, nested in an old tree along the drive. Unsure whether the nest was occupied I tapped the tree trunk and a noise akin to 10 car engines started making me take a step back – the nest was occupied and the chicks were expecting a feed! Rosy used the hollow in the tree for a few years until a strong wind blew off a decaying branch and her nest was exposed with eggs sitting cosily in it. The gaffer tape was collected from the workshop and the branch strapped back in place to last until the eggs hatched and the chicks flew away. The rickety repairs obviously didn't meet the standard set by Rosy or her mate and she no longer nests there.

Occasionally, migrating ibis use the trees for a rest as they pass through, staying for a few days to feed and regain their energy. During the covid restrictions one large dead box tree was covered in ibis, all socially distanced a few feet apart and silhouetted against the rising sun – a memorable sight during a forgettable period of our lives ruled by demagogues.

Permanent residents around the houses and in the olive grove are the superb fairy wrens. They spend the whole day hopping in search of food to sustain their tiny bodies and sit preening and cuddling on rails in front of the windows where they frequent some strategically placed mirrors. Christopher wren is a keen gardener, always in attendance when a sod is turned, knowing that the tread of a foot or newly turned soil will disturb some tasty morsel.

Other small birds which visit occasionally are the diadem firetails and the white-fronted chats. The numbers depend very much on the amount of vegetation there is to camouflage their nests. Pardalotes with their incessant cheeping and swallows annually build their mud nests in the sheds.

The pallid shrike thrush has a mundane name which is an injustice to its clever mimicry and musical repertoire. Nesting in the welding helmets and the children's bicycle helmets in the garage this bird is definitely a morning creature greeting me as I go to work each day and regaling me outside the bedroom window if I have a sleep in. There are a few individuals of this species dotted around the farm calling to each other with well-rehearsed melodies.

In the trees along the main road abutting the front of the farm are the noisy mynas which feed on the insects that are killed by passing motor vehicles. They are noisy too with a busy high-pitched conversation akin to cocktail party chatter. Recently their place as the only member of the myna family has been gazumped by the arrival of the Indian myna. First a pair arrived, quickly multiplying over a few years to a score or more which flock and when breeding seem to take over the nesting spaces of other species. Their arrival is a reminder that no animal has exclusivity in its territory, challengers will arrive and if successful displace a species or reduce its number. I suspect the omnivorous nature and adaptability of the Indian myna has helped it migrate to the country areas from the cities which it has successfully populated. This could humorously be regarded as the avian equivalent of a tree change!

When we approached the local conservation officers to help identify the flora and fauna on the property there were some exciting revelations. There was a remnant copse of silver banksias, another of red box eucalypts, and some rare spider orchids. All these were in an area of the farm which was a combination of remnant forest and scrub regrowth. Flitting through the scrub bushes babbling away were the babblers, a communal bird with excess energy and endearing habits. The species was identified as the white-browed babbler, not as rare as the grey-crowned babbler, but threatened enough to be a reason for us acquiring Land for Wildlife status. We received a government grant to fence off the area of remnant forest and scrub and only native species have access. The result is the babblers have bred to at least two flocks and we are able to observe their antics regularly. They build a communal nest of sticks which when completed looks like a Tina Turner hairdo. Being very shy they will desert a nest under construction if observed by us humans.

They forage as a group babbling away and flipping over leaves and bark to feed on some poor unsuspecting creature which thought it was safe in its refuge. The birds also like a dust bath and will emerge from a puff of dust flapping off the excess and continue to join their mates in the leaf flipping exercise.

Land for Wildlife status requires that we preserve the habitat and look after the species which have been identified as rare or threatened. We have a sign on the front gate heralding the conservation status which attracts some comment from local farmers about blow-in tree huggers. Proud of our efforts in conservation we do not resile from our activities and do not proselytise. We hope that our example will show others that there is a viable alternative to the 'if it moves shoot it, if it doesn't burn it' mantra.

Despite the clarion calls for ecological responsibility there is no recognition of our conservation efforts on a portion of the farm, which consequently is removed from agricultural production to generate income. In the farm valuation the area is deemed to have no value and there is no dispensation in the rating system to reward the conservation initiative. The benefit to us is the delight in observing the flora and fauna in its seasonal and behavioural cycles. Perhaps in time there will be a value assigned to conservation activity and a city dweller who lives in an area of vegetation suppressing end to end concrete will be able to contribute directly to our farm income by buying carbon sink offsets for their contribution to global pollution and habitat reduction. Again I will take aim at the greenies who live in the inner cities cruising in their diesel guzzling SUVs and embracing policies which dictate that others live lives and carry on business in ways that mitigate any contribution to climate disruption. Also known as Nimbies, or 'Not In My Back Yarders', these self-acclaimed climate warriors jet away to overseas trips paid for with money made in concrete citadels which scientists claim contribute 40% of global carbon emissions. They promulgate taxing the farts of sheep and cattle living in the carbon sink open spaces of the rural areas to deflect attention from their own carbon largesse and obtain some sort of collective absolution from their climate confessional. On weekends they uncover their gas-fired barbeques in their backyards to grill steaks and lamb backstraps while complaining about the price. A case of taxing flatulence to sustain opulence?

An aggressive and territorial resident of the farm is the willie wagtail, a member of the fantail family. Sitting with its tail twitching from side to side like a perpetual metronome it will launch an attack on anything that encroaches on its domain – even goannas. Adept at making forays at its target it irritates them until they submit and leave the area. Small, black, white chested with a white eye brows, wagtails are fearless and even let us know our presence is too close to a nest. They build cup-shaped nests made of small twigs tightly sewn together with spiderwebs in a manner which would make a milliner proud. On a moonlit night they can be heard tweeting incessantly on some nocturnal social media platform where repetition must earn viral likes and smiling emojis.

To humour the wagtails and others we have rethought our pruning management in the grove which is the preferred breeding location for many species. Previously we pruned alternate blocks of the seven block grove in spring each year which meant that the birds had their established territory denuded and relatively unsafe for breeding. Now we prune alternate rows in each block leaving adequate well vegetated trees to provide cover for nests and saving the birds from looking for new domains each year. We are not sure that the less disruptive cycle suits the birds but it certainly makes us feel better.

Whenever we work the olive trees, pruning, suckering, or harvesting we observe many nests of different architecture, from rudimentary to intricate. The masked wood swallows, superb fairy wrens, bronze wings, crested pigeons, peaceful doves, and wagtails nest in the trees. Scrub wrens, and other small birds nest in the grass or rows of prunings piled between the rows of trees. Quails keep to the ground occasionally being flushed out by ambling humans.

Every year there is a celebration when the flame, scarlet and red capped robins return to the grove for a brief visit. We do not see them around the rest of the farm. While studying bird behaviour at university I marvelled at the navigation aids used by these and other migratory birds. The knowledge has made me sensitive to major changes in their destination confusing them and perhaps causing them to find another habitat. Seeing them return each year endorses our approach to change as little as possible unless absolutely necessary.

The olive grove is a haven for birds and their chatter and calls fill the days and moonlit nights. We observe them in a non-scientific way, their songs, calls, behaviour, and character being more important to us than taxonomy, identification, and lists. We get anxious when there is silence, thinking that some atmospheric change has warned the birds of an impending storm. We note shrill alarm calls and look for the cause, most often a raptor circling or hovering close by, or a human walking to work.

A major impact of our farming activities has been to provide a new and abundant food source for the birds, wallabies, and kangaroos. As well as the increase in insects and spiders, the olive fruit has become a tasty morsel for a conspiracy of ravens, more than those occupying the Tower of London. Their preference is for the ripe black olives and unlike humans for whom the bigger the better, they prefer the smaller Ligurian style olives. They will strip a tree in a day and knock loosely held fruit to the ground where the corellas, cockatoos and galahs will clean up by cracking the pips and eating the kernels. As we rarely harvest all the olives with some getting frosted and others too sparse or small, it is a tidy arrangement which cleans the tree before the next season fruit set starts. There is some collateral damage when they break off small branches to get to fruit in the thicker vegetation, however this serves the purpose of opening up these areas which is a tactic to reduce infestation by some insects and fungal attacks. The rosellas and wood swallows enjoy a regular meal of olives and unsurprisingly the magpies, who will add anything to their menu, have added olives. Even the wrens have benefitted from a more diverse diet. During harvest when the utility loaded with crates arrives at the house for a lunch break, the shrill summons to lunch is issued by one of the wrens and in a tick they are all over the crates containing the harvested olives, feeding on the array of insects and spiders. They seem to hang out for this daily bonus meal and issue an admonishing 'cheep' when the ute arrives empty-crated. Now they have taken to eating morsels of ripe olives which may have spilled from the crates or fallen into the ute tray as we drive through the grove when harvest is over.

We will cover the kangaroos in great detail later. Suffice it to say here that there are many kangaroos which like the olive grove. Initially we dismissed the tales of other producers that kangaroos eat olives, then after about 20 years we noticed one female pluck some off a tree. More recently we divined that they have collectively taken a fancy to ripe black olives as their usually khaki fibrous faeces have become black and more like a black olive tapenade. In recent years we have not harvested small olives or those which are diseased so, for the first time, the trees have fruit throughout the year available for the roos. The only conclusion is that there has been a change in their diet! Perhaps at a time when the favoured grass is sparse in a drought they tried olives to relieve their hunger pangs and now have shifted their preferences to include them as a regular food. We were never under any illusion that wallabies don't eat olives. We regularly catch them in the grove feeding on green olives, then black olives and when there is no fruit they eat the leaves. They are discerning as their preference is for the kalamata variety which is the most popular Greek olive globally!

Water, Water Somewhere

We have a large dam on the farm and this is home to the waterbirds. There are the permanent or regular residents and a few who are passing through or have possibly lost their way. Prior to the twelve year drought starting in 1997 we had a resident pair of black swans which nested and produced four cygnets. They were charming with one always trailing the others with a slightly crooked neck. Alas they moved on as the dam dried out and we haven't seen them since. A passing flock of pelicans stopped over for a few days, probably depleted the dam of its meagre fish, yabby, and frog population before moving on to more expansive stretches of water with abundant food. We even had a single seagull which was either very lost or on its way to or from a sojourn on the inland briny around the Murray River. It soon flew away after a short stay, probably finding the waveless fresh dam water not to its liking, or it missed the opportunity to steal potato chips from unwary tourists from the city visiting the riverside resorts. Spoonbills grace us with short visits, standing tall with their glossy white feathers and long bill flattened at the end as if a blacksmith has shaped it in a forge.

The permanent residents are the ducks, mainly teal, mallards and wood ducks. The numbers vary seasonally and many breed and muster their many chicks with throaty calls when danger lurks. The brave parents will break from the cover, where their offspring are hidden, feigning a broken wing as they try to lure the predator from the defenceless youngsters. Larger brightly coloured mountain ducks often quack their way across the skies in pairs alighting on the dam. When there is plenty of water they have been known to stay and breed, however they only grace us with their presence for short periods. Then there are the divers which spend their time bobbing above and below the surface as they feed and escape unwanted attention.

The predators for these creatures of the water are duck shooters, by far the most dangerous and indiscriminate in their catch, foxes, and larger raptors. We don't welcome any shooting on the farm but occasionally find shotgun cartridges ejected from the guns of poachers. Our weekends are frequently spoilt by volleys of gunshots from high-powered rifles and shotguns as city people visiting the country 'blocks' vent their blood lust on kangaroos, wallabies, rabbits, and ducks. On occasion we have had to despatch wounded animals from these forays, which include neighbours culling by spotlight on their own properties. Driving to the source of the shooting we have on several occasions heard the crack of a bullet going overhead, an unmistakeable and chilling sound for someone who has been shot at in anger in a war.

The first Australian bird in my life was the kookaburra, brought to my attention through the Australian nursery rhyme 'Kookaburra sits in the old gum tree; Merry, merry king of the bush is he', penned by Marion Sinclair, a music teacher at Toorak College in Melbourne in 1932. A member of the kingfisher family, this medium sized bird has a call which is quintessentially Australian and appropriately named from a word 'guuguubarra' in the Aboriginal language of the native Wuradjuri people from central New South Wales. In modern day recognition of gender equality the rhyme could well be 'Kookaburra sits in the old gum tree; Merry, merry queen of the bush is she'!

To continue the exposures to the 'burra' in early life we gave our grandson a toy replica which had a motion activated kookaburra call. Living in London under the flightpath for Heathrow where an aircraft goes overhead every two minutes, our son discovered the noise of the jets activated the call – a very satisfactory outcome! Alas the motion sensitivity was disabled so the family could get some uninterrupted sleep. Kookaburras feed on small creatures such as mice, snakes, lizards, frogs, and insects. Despite their classification as kingfishers they rarely eat fish. Their call often cuts through all others, mainly early in the morning or evening. There are several who have made the farm their territory, however we have not seen them in the olive grove until recently. Competing with the kookaburra for riverine inhabitants as food are the lanky grey herons which are seen on the farm dams from time to time. The shags or cormorants feed on any fish and yabbies, the fresh water crayfish that are in all the dams. The shags enjoy the tall dead trees which border the dam where they can be seen spreading their wings to dry before another fishing expedition. Their solitude when scanning for prey alone on a tree or rock gave rise to the expression 'like a shag on a rock'. All these birds which feed on frogs, fish, and yabbies are a barometer for the ecological health of the dams with no chemicals used in the catchments.

Before we explore the other inhabitants of the water and the importance of water to the ecosystem, there are a few other birds to mention. The boobook owl is seen around the sheds where there are often mice to feed on, and a young one lived in the olive grove for a short while before moving on to establish its territory for its adult life. The plovers or lapwings nest on the ground around the dam and diligently protect their exposed nests from roaming sheep or foxes. Their choice of location for the nest is not the most sensible and consequently we rarely see fluffy, young black lapwings as the eggs or chicks are easy prey for goannas, foxes, and winged predators. This year the resident pair have almost managed to rear two young. For some reason they hang around the irrigation pump on the large dam. Observing one of the chicks taking cover under the walkway, I looked for it escaping as I walked to the pump. It disappeared so I searched further. Almost impossible to see, it had stretched itself along a grey branch under the walk and blended in with the grain of the weatherbeaten wood. How did it learn to do this? Did the parents, which were chattering away, tell it what to do? I was impressed and happy to see that with this learned or instinctive self-preservation the two chicks are now adolescent with a good chance of reaching adulthood.

Mudlarks which look like small magpies also live around the water as their name would suggest. Small waders patrol the muddy edge of the dam stepping lightly with regular pecks into the shallow water to extract a snail, worm, or other morsel. The multitude of imprints of the small birds patrolling the pristine mud along the edge of the dam look like the flight radar images of the aircraft on their approach to Heathrow!

The four different ecosystems on the farm, the remnant forest and scrub, the open pastures, the dams, and the olive grove, support many species of birds and insects, the numbers of which ebb and flow with the annual seasons and the climate variation from wet years to drought and everything in between. There are many species which will not be specifically mentioned here which are integral to the food chain. No doubt as we continue to observe the wildlife we will discover and appreciate more of their contribution.

The large dam which we built is a constant source of pleasure. It provides water to irrigate the olive grove, quick access to helicopters and fire trucks to fight local fires, and a kaleidoscope of reflections and colours which impart a peace of mind and serenity to us humans. In its early days we stocked it with redfin fish and held weekend fishing parties where many a child learned to fish. Redfin are great to eat and the day's catch ended up on the barbecue. These days helped keep the population down and unfortunately ended when a lengthy drought caused the dam to dry out to a screed of cracked mud. We never restocked it as the laws relating to stocking dams with redfin, also known as English Perch, changed to protect native fish in downstream lakes.

One day we noticed shoals of little fish in the dam which had been replenished by good spring rains after the long drought. Excited, we thought that the redfin had somehow survived the drought with roe buried deep in the moisture below the cracked mud surface. The fish grew to a substantial size over the next two years before the next drought. The cormorant and heron populations increased and fed off the abundant supply of fish. When the dam started to dry out we noticed that there were some large fish trapped in the bottom part of the dam – the dam being divided in two. We irrigate from the bottom half and keep the top half in reserve. Wanting to save the 'redfin' we spent an exhausting afternoon catching the sizeable fish and carting them in buckets about 30 metres to the water in the top half. The bottom of the dam was muddy and our feet got stuck, the fish were slippery and not keen to be caught even though there was little water and they were floundering in the mud. Eventually we transferred about 15 and retired for some welcome beers, well pleased with the rescue effort. Showing the pictures of the fish to a keen fishermen friend he informed us with a smile that the fish were not redfin but carp which are considered the pariah of the rivers. We could do nothing but laugh and eventually as the top half of the dam dried out the fish were caught and devoured by foxes and raptors. The roe of the carp must have been brought in by birds from local reservoirs or rivers. There have not been any fish in the dam since this episode.

The common yabby inhabits all of our dams and from time to time we go yabbying – catching these freshwater crustaceans in drag nets. They grow to a reasonable size and are delicious to eat. Any small ones are tossed back in the water, as are females which have eggs attached. When caught from the muddy dams the yabbies are placed in fresh water for a few days so the muddy water is flushed out of their system. Cooked the same way as crayfish the flesh is sweet, and enough can be caught in one expedition to feed three or four diners. Specially designed traps baited with rotten rabbit or some other delicacy are an efficient and lazy way to catch large numbers of yabbies. The design of the traps is regulated to ensure that the freshwater turtles are not caught and drowned. Eastern long-necked turtles sometimes visit the dams and occasionally we find them on dry land travelling between dams. They are mainly carnivorous and will eat insects, worms, snails, and even mosquito larvae. Like yabbies they can survive by burrowing into the mud and remaining dormant for extended periods when dams or water courses dry up, and emerge when the rains refill their habitat. They breed by laying numerous eggs in moist soil then covering them.

Every year we hear during the summer rains a unique sound, the phonetic 'pobblebonk', emanating from all the dams and billabongs. It builds to a stereophonic orchestra of many tones and inflections as the eastern banjo frogs or pobblebonks play their symphony for procreation. Pockets of slimy foam resulting from mass spawning subsequently punctuate the surface of the dam. We love the sound of the pobblebonks as another testament to the natural ecosystem we have protected with its chemical free status. In our enthusiasm as citizen scientists we joined Frogwatch and diligently recorded the frog calls at the big dam after dark, sending the results off to the administrators for identification of the species from their calls and croaks. The excitement of the first reports soon waned as repeated recordings revealed the same species and the effort of heading off to sit by the dam in silence at various hours of the night lost its gloss. However, for those with a dam close to the house where the recordings can be taken while sitting on a veranda cradling a glass of red or sipping a good rye, we recommend citizen science! Given a huge backlog of remote video cameras recording the movement of wildlife, some genius came up with the idea of engaging the retired population with computers to assist in going through the video recordings, hours of tedious monotony, to detect anything of interest. This gave everyone a positive result, the scientists had the work done, the retirees felt they contributed and engaged with nature. What a brilliant idea!

The Guardian newspaper biennial bird of the year run in Australia similarly engages with the population to select their favourites and at the same time getting across important information about the species which are endangered. The Fat Bear Week in Katmai National Park and Preserve in Alaska celebrates the brown bears stacking on the fat by eating salmon to carry them through the winter. The online competition attracts over a million voters worldwide who select the Champion Fat Bear from the listed contestants. Who will win, Chunk or the blonde female Holly, Grazer, or the two-time champion Beadnose? Last year it was the appropriately named Grazer!

The place of the brown bear in the ecosystem is established with this empathetic championship far more effectively than the slogans of environmental warriors. The involvement of citizens in observing nature serves the dual purpose of engaging them to appreciate the natural ecosystem and using them as cost-free data gatherers and interpreters.

Water is essential for life and often in short supply on the farm. The summers are dry and the spring rains are variable with no guarantee that the water table will be replenished and the creeks will run to fill the dams. The flora and fauna have adapted to ensure survival during long periods of drought and quick exploitation of short periods of high rainfall. During the summer and periods of dry weather we irrigate the olive trees in the grove and periodically have to purchase fresh water for the house water supply. We have birdbaths in the gardens and delight in watching the birds routinely come in at certain times of day to bath and drink their fill of fresh water. The bees also come in to collect their share of liquid, interestingly preferring the water from the dam to the fresh cleaner water from the tanks where we collect rainwater from the roofs.

There are, during the summer, sessions of frenzied bathing activity involving the honeyeaters of which there are many species. They generally come in during the heat in specific groups, the most common being the white-plumed and yellow-tufted honeyeaters, smaller groups also visit comprising the strikingly decorated new holland honeyeater, yellow-faced, white eared, yellow-plumed, black-chinned, white-naped, and painted honeyeaters. The naming of the honeyeater family surely tested the repertoire of the taxonomists reducing the descriptions to such boring prosaic names as the 'brown honeyeater'. The daily bath time is akin to children jumping into a swimming pool and squealing and splashing in the process. More than 30 honeyeaters have been counted on the rim of the birdbath at one time with many waiting their turn to join the fray when another flies off into the tree to dry. Collectively known as 'the splashpots' they will look forlornly towards the house if the birdbath has not been filled and shame the occupants into carrying buckets of water to replenish the depleted pool.

In the middle of any of the water baths or drinking bowls we place stones or rocks, or in deeper containers sticks, to assist any small birds which fall into the water. Too often we found small birds such as wrens drowned and the 'islands' and 'ladders' have been effective in eliminating this cause of death.

In the evening a flight of galahs comes in and rings the edge of the water bowl, taking turns to bob up and down for a drink. In the late afternoon sunlight they paint a picture of pink and grey highlighted by the setting sun. These are the serene pleasures of providing habitat and food for these beautiful creatures and indulging in the privilege of observing their behaviour in the natural environment. The galahs always roost in the tall box trees around the houses, sometimes disturbed and no doubt irritated by the late night pedestrian traffic as dinner guests return home.

During the dry season all the creatures crave water and shallow water bowls are dotted around to keep them hydrated. The reptiles such as the eastern shingleback, stumpy tail or blue tongued skink frequent the watering bowls. Their scales look like those of a snake and have given many frights when almost underfoot. The water sometimes attracts the lethally venomous brown snakes which we do not tolerate around the houses. Instead of killing them we relocate the water to areas away from the houses and the threat is removed. The well camouflaged goannas live in holes and dead trees appearing infrequently to avail themselves of water. Their presence is usually heralded by bird alarm calls, especially in the breeding season.

One very dry year we noticed an old wallaby called Culvert up in the grove – named because he made a culvert in the road his home. Sensing his distress from thirst we took a bucket of water up to him and he was so grateful he virtually climbed into the bucket to slake his thirst. Thereafter we always ensured that there was water available for him either through the irrigation system or by special delivery.

The summers are hot and long and the onset of autumn is welcomed by all on the farm. Everything cools down and prepares for the coming winter which can be cold with temperatures well below zero. The farm becomes green and the harshness of the summer dry is forgotten in anticipation of the songs and colours of spring as the earth warms up and the flowers decorate the landscape. The four seasons are as melodic as Antonio Vivaldi's concerti, the difference being that the natural composition varies from year to year.

Dragons, Snakes and Bladders

The reptile population has grown exponentially in the last 30 years. The shinglebacks or stumpy tails are ever present during the summer, methodically making their way around the feeding places. They are omnivorous, generally feeding on insects and snails, supplementing this fare with opportunistic feasts of the yellow flowers of cape weed which emerge in abundance in spring. The 'stumpies' move from plant to plant devouring the flowers with an audible pop as each is removed from its stem before moving on to the next one. Like all the lizards, the shinglebacks hibernate and emerge to bake in the sun in spring before heading off in search of the spring flowers which supplement their insect diet.

Many times when we are quietly going about or business of pruning the olive trees there will be a frightened yell – a mixture of fear followed by relief that the bark-like protrusion on a branch is only a bearded dragon and luckily has not been mutilated with the pruning shears. For their size the dragons are fearsome looking reptiles and can move very rapidly along the ground or up a branch. One unexplained phenomenon which is occasionally observed is the dragons all climbing up to exposed areas of trees or fence posts and facing north – like the figurehead on the prow of ancient sailing ships.

The largest lizards are the goannas, the longest on the farm is named Kerry. Often mistaken for snakes when only parts of the body are visible, they are adept climbers and will raid bird's nests for meals of eggs and chicks.

The resident species of snake on the farm is the eastern brown snake, the second most venomous land-based snake in the world and responsible for a high proportion of deaths from snake bites in Australia. We treat them with respect, sometimes with adrenaline fuelled fear, and generally try to avoid them by understanding their movements and favourite habitats. There have been a few confrontations, only one resulting in the death of the serpent – it was either it or me. In the sheds from time to time there is a population of mice, a favourite food of the brown snake. One day a scaled tail of reasonable size was seen disappearing under a board, the only conclusion was that it was a large brown snake. Needing to enter the shed we attempted to lift the board from a safe distance to encourage the snake to go down its hole or head off elsewhere. A species known for its aggression, the brown emerged and managed to strike at my leg as I jumped backwards onto the utility tray which was three feet off the ground. An Olympian feat which could only be the result of a massive shot of adrenaline. Fortunately for me the strike was prevented from penetrating my skin by the thick roll of my sock. Now I know the brown snake is a protected species but a fine or jail seemed a much better option than 'venom-induced consumption coagulopathy, serious systemic envenoming including hypotension and collapse, thrombotic microangiopathy, severe haemorrhage, and cardiac arrest. Plus nausea and vomiting, diaphoresis (sweating), abdominal pain, acute kidney injury and seizures.' Apparently a headache can develop in 15 minutes and clotting abnormalities within 30 minutes. Since it would take at least an hour for an ambulance to get me to a hospital, where there would no guarantee they have the antidote, it had to be the snake or me. Believe me my adrenaline accelerated brain came to this conclusion in a nanosecond and still in super-active mode I assaulted the reptile with a shovel wielded with primal frenzy, and eventually despatched it. It fought back rightly fearing for its life and left its assailant in a trembling state of shock.

So you wonder why we don't shoot all snakes on sight. Firstly, we don't always have a shotgun to hand and even if we did we often could not use it in confined spaces like sheds or houses. Secondly, they are part of the ecosystem and protected, they have a place and our approach to all creatures which are a threat or pest is to try to understand them and live with them. We have had a few subsequent encounters with brown snakes. A while back we had the pastures mowed and the hay baled to use as mulch on the young olive trees. The contractor moved the bales to the grove and stored them in neat rows ready for us to unroll on either side of the trees in spring. We did this by pushing an iron bar through the middle of the bale and pulling it with the ute so it unrolled as required. Easy exercise, except where there is dried grass seed there are mice, where there are mice there are snakes – or on this occasion one large brown snake. Given our previous experiences, we were alerted to its possible presence by the mouse tunnels when we moved the first bale. No matter how prepared one is for a snake it is always a scary confronting experience. The snake appeared under the second roll and the person thrusting the metal rod through the bail alighted onto the bonnet of the vehicle with a huge exhalation of breath. So we knew the snake was there, we weren't going to kill it and needed to finish the job. We decided to go and collect the next bale from the other end of the line – the snake had the same idea and we met again. Using pitchforks and working from a distance we encouraged the snake to move elsewhere which it did and gingerly we extracted the rest of the bales unsure whether the snake had a mate which was ready to give us another adrenaline rush.

We see snakes from a distance from time to time and note their movements. On two occasions a sixth sense has stopped me from stepping on one. The first was in early spring when walking through the bush I stopped in mid-step to see a brown snake curled up in the sun at the entrance to its hole. I still shiver at the thought of stepping on it – luckily it was cold and the reptile hadn't warmed its blood up enough to be quick in its reaction. Another time I was riding through the bush on the three-wheeler motorcycle and for some reason stopped, looking down to the ground I saw the head of a large brown snake just in front of the front wheel. It just lay there and looked at me. There is no reverse on a three-wheeler, but I found reverse and got the hell out of there.

The final major episode with a 'brown' was my son's experience while pulling up irrigation lines which were buried under long grass in the olive grove. He pulled up a length and disturbed a snake which he says was lying across it. He jumped back and took flight with a yell and informed us of the snake. As there were other workers in the vicinity it was important to know which direction the reptile had escaped so I asked him where it had gone. He responded with several expletives which showed that he didn't know and didn't care. We moved to work in another part of the grove!

Thankfully our meetings with snakes are infrequent, however as you can see from the above tales, they are memorable and adrenaline fuelled.

The prey of snakes, goannas and some birds are the small lizards which we disturb when suckering olive trees or moving dead logs. Their presence and that of other critters in logs, old fenceposts or piles of dead vegetation has stopped us from burning them in tidying up the farm. We now see everything as habitat and while enjoying the neatness of a manicured farm realise that untidiness in the eyes of humans must be tolerated in the interest of encouraging the fauna. Outside their own nests and immediate surrounds we doubt that creatures see as high priority the reordering of perceived chaos of their environment.

Galahs sitting in a dead gum tree.

Trannie, the old wallaby eating his daily treat.

Cassius, the large male kangaroo in the shade under the olive trees.

Zeus, the wedgetail eagle.

Bath time for babblers.

A cockatoo doing acrobatics hanging from a tree.

Cockatoos in the almond tree.

A male kangaroo defending his territory.

The love train (echidnas.)

Green combed spider orchid.

A bed of sun orchids.

Red spider orchid.

Dirt, Mud and Dust

Without the soil there would be no farm. It is the foundation on which the ecosystem is built and part of the ecosystem itself. It is alive, full of microorganisms and larger inhabitants which nurture, aerate and determine what will grow in and on it. Its profile is a result of millennia of erosion, a product of the seasons, leeched by wet years, desiccated in dry years, always resilient and growing vegetation. Even when desertified it slowly rebuilds, cell by cell, microbes to microflora, and very ancient plants to massive eucalyptus trees.

There are very clever, seemingly mundane, things that happen to the soil which perpetuate and protect the flora and fauna. For example, the deciduous trees drop their leaves around the base of the tree and then they decompose with the help of insects, microfauna, and fungi to be rendered back to the nutrients which the tree then recycles. Dust blows hitting the tree trunk falling around the base giving a layer of topsoil bringing new nutrients to sustain the tree.

Trees grow in copses or groups to protect each other from strong winds which topple individual trees standing alone in a paddock. Branches which break off trees fall to be rendered by termites, fungi, and other microorganisms to nutrients. The tree canopy shades the root area to reduce evaporation of precious moisture and the wind which accelerates loss of moisture is broken up by groups of trees creating a windbreak and again preserving moisture.

Not being a soil scientist, it is difficult to detail the ecology of the soil beyond some general observations which will suffice here. Given our approach of not applying any artificial fertilisers or tilling the soil we rely on natural means for conditioning the soil to improve it to grow our agricultural outputs, the main one being the table olives from the olive trees. Life in the soil takes us away from the creatures and plants we can see with the naked eye to those which can only be seen with microscopes in laboratories. We have no microscopes and there is no laboratory within hundreds of kilometres, and even those are expensive to access.

The soil over the farm varies but is in general what farmers call 'marginal', a euphemism for not very fertile. The vegetation which grows in it is the first indicator of the ability of the soil to sustain life. The groundwater in many parts of Australia is saline and the farm is no exception. Areas which were denuded of trees needed to suck out the groundwater through transpiration became saline as the water table came to the surface, killing all vegetation and the crystals of salt could be seen in striations glistening in the sun. One of our first tasks was to obtain some salt resistant red gums from a farmer friend close to a saline lake. We planted these in the areas where the groundwater came to the surface, called discharge areas, and over decades the eucalypts have thrived and the surface salinity has receded. With the recession of the salt the groundcover has returned and bound the soil limiting gutter and sheet erosion. We have allowed the trees to grow naturally along the gullies and this has reduced the erosion and provided habitat while acting as windbreaks from the hot northerly winds in the summer.

The soils on the farm are generally described as sodic which means they are acidic with poor soil structure inhibiting root penetration and drainage. If a hole is dug to plant a tree, after many years the tree will have grown little as the roots are bound inside the original hole unable to penetrate the surrounding soil. The poor drainage can lead to waterlogging of trees which are attacked by root diseases. Olive trees like alkaline, well drained soils and are prone to waterlogging. Knowing this we undertook soil surveys to locate the best area for the grove and, prior to planting the trees, ripped a metre wide strip so the soil was broken up to give the roots a chance to spread. We also mounded the topsoil along these strips to provide more drainage. Soil compaction limits the penetration of roots between the rows where heavy vehicles drive and cloven hoofed animals such as sheep and cattle also compact the soil where they roam. After the initial work of ripping and mounding we have not used heavy machinery, doing all the work of harvesting and pruning by hand and only using a utility and trailer for any carting. We do not apply any chemicals for pest or weed control, or fertiliser, so do not use heavy spraying equipment and livestock does not have access to the grove. With the light utility or truck we only drive down every second row so every row of trees has access to an area where there is no compaction.

I can hear the cynics saying, 'how do you control the weeds and reduce the fire risk in summer without tilling the grove'? Perhaps they have visited the vast groves in southern Spain where every suggestion of vegetation is either removed by tillage or application of herbicide. Branches from pruning are burned and a bare earth policy is applied. As a result erosion is rife and there is no natural fertilisation from the breakdown of vegetation from the trees. We do the opposite, we let the grass and weeds grow and rely on the 'mowers' or resident mob of kangaroos to graze and keep the grove relatively tidy. Being soft-footed the 'roos' don't compact the soil to any degree and do little damage to the trees. We will talk about the kangaroo population in its own chapter later, here we acknowledge their contribution to vegetation use in the grove and the benefit of the manure they spread around. The noxious weeds they don't eat we chip out using hoes and if we keep on top of the few we have there is no need to resort to weedicides. New seed blows in over the fence, is carried by birds or is spread when the farmers agisting their sheep on the property bring in hay or grain from sources outside their own operation. We have to be constantly vigilant to eliminate any new incursion of unwanted plants before they take hold.

Allowing weeds to grow and insects to thrive is a means of aerating the soil where roots penetrate the clay and ants move subsoil to the surface as they build the underground labyrinth of their nests. Earthworms are good too with their processing of organic matter and leaving casts which improve soil fertility. Their presence is an indicator of the health of the soil.

The birds help with some tillage. The galahs, cockatoos and corellas will work away at digging out the onion grass with a flock of them turning over a few square metres of the very thin layer of topsoil in a day. Research quantifies this activity revealing that a cockatoo can consume 200 onion grass corms an hour with a flock accounting for tens of thousands in the same time. Some of the galah pairs have specific allotments which they return to regularly to expand the area of digging. Respecting their majestic flight we have named two pairs Rolls and Royce and Pratt and Whitney after the leaders in jet engine development. They have allotments up near the olive grove and receive our felicitations when we travel to and from work. I am sure that this sounds silly to many, to us it is like passing friends on the street when taking the kids to school or heading off to work in the office each day. They don't respond as they are very intense in their work of gathering the daily food.

When suckering the olive trees we find little caches of clover burr secreted in the crevices in the bowl of the trees or even up the trunk in the cups formed by branches. We are not sure who is squirrelling away the burr or even why. Presumably they eat the seeds and we suspect either birds or mice. To collect them there must be some digging adding to the scarifying of the soil surface. In this and other ways the nitrogen fixing clover plants are spread around and add to the natural fertilisation of the soil.

The soil microfauna and flora is a complex population involving symbiotic, parasitic, and independent fungi, nematodes, annelids, and bacteria, all of which add to the ability of the soil to digest and recycle vegetative matter offering minerals and other nutrients to the trees. We have seen an increase in earthworms which play an important role in increasing the fertility of the soil and bringing it closer to the friable loam which is so much better for everything than the sodic clay. The magpies and ibis with their long beaks extract the earthworms for food and in doing so help aerate the soil.

Home to a range of fungi and bacteria which can attack or affect the roots of a tree, the status of the soil is very important and the tree canopy is often the first indicator of something amiss. For example, dieback of the of a small branch, or in some cases larger areas of the tree, can be an indication of the multiplication in early spring as the soil warms up of the spores of the verticillium fungus or phytophthora which attack the young fresh roots being spread out by the tree. Usually associated with wet seasons and waterlogging there is little we can do to combat these pathogens other than improve the drainage of the soil and encourage the vigour of the trees.

During the summer the hot western winds blowing from the red arid centre of Australia often carry a heavy load of dust from the desertified outback. Sometimes so thick that one can barely see 100 metres, it covers the buildings and ground with a patina of dust which we regard as free topsoil. Covering the ground with binding vegetation ensures it isn't our dust which is carried away, we try and break up the wind so the dust is deposited and held around our property. One can see the slightly elevated soil level along fences and the ridgelines in the olive grove where over years we have successfully captured the free topsoil deposits.

Without the soil there would be no growth and we need to be careful that we do not disregard the flora and fauna which may be out of sight but should never be out of mind. Careful observation and interpretation of all things that grow or inhabit the habitat above ground will give many clues on the health of the growing substrate and remedial action can be taken to limit damage where it occurs.

Sometimes human intervention can threaten the natural status of the soil. The farm is on the fringe of the 'golden triangle' in north central Victoria, an area which attracts the attention of gold fossickers and more serious international mining conglomerates. We have little control over the mineral rights bar a few metres close to the surface and cannot stop prospectors from exploration. We rely on governments, always a dangerous situation, to enact legislation to protect the environment. There is tension between the flow of mining royalties into often depleted state coffers and the social responsibility of preserving vulnerable ecosystems.

One large mining company assessed the area to determine the viability of solution mining for gold. Their geological maps showed that there is an ancient river bed under part of the farm, a possible source of alluvial gold. Their idea was to pump a lixiviant solution into the riverbed upstream and then suck it out downstream after it had reacted with the gold to form soluble salts from which the gold could be extracted. The idea of the groundwater on the farm being infused with some toxic chemical caused considerable concern. A hole was drilled at a high point on the farm and samples taken. The results of the assay lie buried in the archives of the Department of Mines in Ballarat and long may they remain there! We were paid a nominal amount for access to the drill site and learned that the underground water was too saline for us to use for irrigation should we decide to sink a borehole. Fortunately solution mining was banned and we were spared a battle with the mining companies. Somewhere in the olive grove a capped and lined drill hole remains as testament to this event, we could get the GPS reference if we wanted to access it – but we don't!

Fermenters, Defenders and Tormentors

Writing and looking for chapter headings, an author frequently searches the internet for a synonym, an explanation, to check on a fact, and all sorts of other information. These searches are akin to those which the burgeoning artificial intelligence robots undertake to come up with an answer to your question. My fear is that these searches without interpretation will maintain the status quo, both because most of the questioners are status quo and the search algorithm will lean towards the heaviest weight of data and opinion which underpins the status quo. For example, in searching for some rhyming chapter title to cover bacteria and other microbes the overwhelming references and synonyms for bacteria were associated with disease, illness, mostly those afflicting humans. There were few, if any, references to the good things that bacteria might be used for – such as making wine, preserving table olives, pickling gherkins. A search for 'good bacteria' reveals how important bacteria and other microbes are in the human biome or human microbial communities which assist in digestion and other metabolic functions. And yes, there are even good viruses which naturally help the body detect cancer cells and subsequently deal with them. As with many organisms which have the ability to multiply very quickly, microbes will take advantage of ideal conditions to mass produce and consequently upset the natural balance of an ecosystem or being.

Macro examples of this are the locust and mouse plagues we have experienced on the farm. Conditions, usually involving the weather and human provided food supply, occur where these insects and rodents multiply to plague proportions and travel in search of food to sustain the hordes produced by a very short breeding cycle. Luckily these plagues are rare and their destructive existence is for short periods. The mice on one occasion carpeted the roads around the sheds where we stored stockfeed and needless to say devoured a fair quantity of the grain stored. Predators had a feast with little impact on the numbers. Grain farmers incurred huge losses as they do when there are locust plagues. The locusts devour all vegetation and leave denuded plants in their wake as they march on to the next food source. Motorists cover the grills of their cars to protect the radiators from being clogged up with dead insects. Eventually either the food runs out or the weather changes to a less compatible temperature and the plagues end. Our most recent relatively localised plague is the olive lace bugs which have decimated abandoned groves nearby denuding the trees to the extent that a few years recovery will be needed before they bear a commercial crop. Our grove is infected but luckily because of our almost daily observation while working in the grove we have been able to forestall major damage – we hope. Why did we get the olive lace bug plague – we think because the scale, which dominated the trees protected by the ants which fed on lace bugs, were devoured by a surge in ladybirds – so the scale died, the ants died, the ladybirds died and the lace bugs said 'aha, it is our turn'. Now they will die because they have destroyed the leaves on which they feed, so the whole cycle will start again! First the scale, then the ants, then the ladybirds, just like the bush recovering after a fire. It is fascinating to observe. Of course we could control the whole natural cycle by spraying on copious amounts of pesticides and fungicides to destroy all life, and by flow-on affect the birds, the echidnas, dragons, other insects which are exclusive pollinators of the orchids – which will also peter out. All in the name of profit!

Spraying with fungicides and bactericides would destroy the microbes on the olive fruit which we use for natural fermentation. Here we will digress a bit to describe the processing of table olives or olive oil where yeasts and bacteria are of great importance, and the natural defences of the olive fruit and tree against them are equally important. While using the olive because we know more about them, all plants will have similar mechanisms which are species specific and manifest themselves in different ways.

If you follow the promotion of extra virgin olive oil to consumers you will be aware that there is huge emphasis on the health benefits. The growing list of human ailments which researchers claim are cured or mitigated by the various phenolic components, or anti-oxidants, in the oil is impressive and includes cancers, cardiovascular disease, dementia, osteoporosis, and heart disease. One would think that the plant has designed itself to be highly desirable to humans so they would nurture it for their own benefit, in other words farm it! Not so, the olive and other trees produce these components to protect themselves from viral, fungal, bacterial, and insect attack. Not only are they found in olives and olive oil, but also in the leaves from which the medicinal olive leaf extract and the herbal olive leaf tea are made. If one squashes a small glossy scarab like black beetle found on the trees, it smells like olive leaf extract sold as a panacea for many ills, evidence that the beetle naturally extracts this 'elixir'. Varieties of trees have different levels of the anti-oxidants which can be profiled in the intensity and complexity of the flavour of extra virgin olive oil. The intensity and range of the phenols gives the extra virgin olive oil protection from degradation through oxidation, with more intense stronger flavoured oils lasting longer than the 'delicate' oils. We see this variation in protection in the grove with the arbequina variety from Spain, which produces a delightfully flavoured milder olive oil, being more prone to attack by sooty mold, which blankets the tree with a soot-like patina, than its Spanish relative picual which produces a more intense oil with the higher level of polyphenols protecting the tree.

As the trees evolve in different environments they will respond to the various microbes which attack them by developing protective compounds which will give them protection or immunity. Microbes mutate so the trees can never rest on their laurels and assume they have perpetual immunity.

In farming them, humans use pesticides, bactericides, and fungicides to protect the trees and fruit so they no longer have to manufacture their own defence mechanisms. As the microbes mutate and develop resistance to the chemicals we apply, someone in a laboratory manufactures a new compound to eliminate the mutant and the cycle starts again. A concerning feature of the trend towards intensification of olive tree farming is that the trees most adapted to hedgerow planting and management are those with lower polyphenol levels, hence lower resistance to disease. The close planting of the trees will facilitate the movement of infections or insect attack and the lack of resistance of the trees will contribute to potentially disastrous vegetative breakdown and commercial loss. In the current environment of blaming everything on climate change, this will probably be added to the list. It is human intervention and manipulation which is the cause, planting trees with low resistance in crowded conditions and relying on massive application of artificial controls to keep the trees productive. This 'scorched earth' approach to olive tree management also removes some of the natural bacteria and yeasts which are specific to a particular variety and grove which defines the different olive oil and table olive products unique to that environment.

In a positive case of human interference we have been able to contribute to in the interest of sustaining a species of indigenous tree relates to a banksia. Near the house dam we have a small copse of the native silver banksia which has virtually disappeared from the area. Ravaged by rabbits in the early years of the farm we actively protected young trees with wire netting tree guards to increase the number. The local conservation and wildlife officers approached us with a request to test the drought resistance of our banksias with a view to cross-breeding them with some from the coastal region which had little resistance to drought and needed to 'harden up'. Not surprisingly, after years of surviving many dry years on the farm our banksias had self-selected and exhibited a genetic profile different from the coastal trees. Seeds were selected and cross-bred trees were propagated and planted in the coastal region to maintain the banksia population. The 'new' trees were offered to us but we did not want to interfere with the genetic stability of our own little group of survivors.

Some of the polyphenols which are desirable in the flavour of olive oil are not perceived as desirable in table olives – especially the bitterness which is imparted by the compound oleuropein. To prepare the table olives for human consumption the bitterness needs to be removed and the fieriness or hot flavour imparted by oleocanthal, an anti-inflammatory, dissipated. The ancient traditional method to do this was to wash the olives daily for 10 days resulting in the water-soluble flavour components, such as oleuropein and oleocanthal, being washed out. The olives would then be preserved with a mixture of vinegar and salt.

The modern method is to ferment the olives using starter solutions of bacteria and yeasts or using the naturally occurring microbes on the olives. The former process gives more control resulting in a standard product while the latter takes longer and because of the seasonal variation of the bacteria and yeast population the final product will vary from year to year. Table olive aficionados will detect the processing method as the natural yeast and bacterial fermentation results in a product with more complex and alluring flavours.

The lactobacillus bacteria naturally present in the olive convert the sugars to lactic acid which is a natural preservative similar to acetic acid or vinegar. The type of fermentation is controlled by the salt level in the fermentation brine, salt levels between 6% and 8% encourage the lactobacillus fermentation and those between 8% and 10% encourage a yeast fermentation. A combination of both gives a full flavoured olive with the sugars being fully utilised and a stable product ready to be stored or consumed. The bitterness and fieriness slowly dissipate as the compounds imparting these flavours are hydrolysed or oxidised.

The point being made here is that the natural microbial population of an ecosystem can be used to produce a naturally fermented product. Like trying to control an entire ecosystem there are two options, one of intervention, standardisation, and control through the application of man-made chemicals and fertilisers, the other of careful monitoring of the ecosystem allowing the flora and fauna to achieve an equilibrium, unmolested. The first option is high cost requiring high yields to make it commercially viable – what I call the maximising approach with high inputs, high commercial risk requiring high yields. The second I call the optimising approach – low inputs, low commercial risk, and low yields. Having worked with both systems, my personal preference is the second option. It is easier on the environment and lifestyle.

We have concentrated on olive trees as an example while knowing that other trees have their own similar mechanisms. During dry periods we watch the eucalypts shed leaves to reduce the transpiration and water usage. Olives do the same when they have a large fruit load, they will first drop leaves, then as it gets drier they will pinch off nutrient supply to some of the olive fruit so it dries and drops off reducing the fruit load to a manageable level. If the dry persists all the fruit will shrivel and drop off and further leaf drop may occur. This is a remarkable system for survival which we disrupt with irrigation and harvesting, all acceptable if managed in concert with the trees natural self-management.

There is no compartmentalisation of the ecosystems on the farm, the soil is connected to the air and all the plants which grow and the creatures which traverse the ground. When considering some agricultural action it is incumbent on us to assess the possible effect at every level and subsequently observe the impact for future reference.

Lichen to Eucalypts

Most of the description so far has related to the section of the farm which we have adapted to be an olive grove to provide us with the income to indulge the minimal commercial or agricultural use of the 'Land for Wildlife' set-aside. The olive grove and the pastures, which take up two-thirds of the farm, are a form of monoculture and do not have the diversity of the fenced-off area where only native animals can roam and sheep are excluded.

Eucalyptus trees are the main family of trees which form the canopy. There are individual trees in the paddocks, rows growing in the gullies and along the roads and a forest of remnant vegetation at the back. The main species are grey and yellow boxes, forest and river red gums, white gums, and long-leaf boxes. Tall and imposing, these trees have wood so dense that driving a nail into dry timber from them is nigh impossible. There is one small copse of red boxes which seems to have been spawned from a very large tree which was felled and left numerous small progeny as a testament to its existence. There are no other red boxes on the farm or in the immediate area. Apparently the wood is greatly sought after for furniture. Prior to our arrival there was some harvesting of the large box trees to yield fence posts which are so hard that they last for over 100 years.

The next layer down comprises she-oaks, also known as bulokes which are members of the casuarina family. The buloke produces the world's hardest wood, 5,060 newtons on the Janka scale. The Janka hardness test, created by Gabriel Janka in the early 20th century, measures the resistance of the wood to denting and wear. By comparison, European olive wood has a Janka of 2,710 newtons and the spotted gum 2,473 newtons. The hardness of the buloke wood resulted in it being used by

Aboriginal people to make weapons such as boomerangs and spears, and fashion numerous tools used in daily life. The trees have nitrogen-fixing nodules on the roots, another example of a symbiotic relationship with good bacteria, and the foliage is drooping and horsehair like. The wind whistles through the foliage of the small clumps of trees in contrast with the whispering of the wind through some poplars we have planted around the grove. During droughts some farmers fell she-oaks to provide feed for sheep. At the back of the farm there is an area of sandy soil which the casuarinas like and it sustains our largest population of these trees. We have noted that they are starting to spread to other areas of the farm and increase the diversity along creek lines.

Attached to many trees is the ubiquitous mistletoe. This hemiparasitic plant hitches a ride on the host via a haustorium which efficiently taps into the water and nutrient supply of the tree. Mostly spread by birds, the infestations on a single tree can be so numerous that during dry periods when the tree vigour is at a low ebb the tree will die. Thanks to the presence of mistletoe we have a population of mistletoe birds. The Lord giveth and taketh away!

Underlying the tall canopy there are layers of understorey which comprise a great diversity of flora which in turn supports a diverse multitude of fauna. The tallest of the bushes are the very ancient looking and spiny hakea. The foliage is uninviting because of its thick spiny nature and one does well to avoid having to deal with them. I note some gardeners grow hakea as ornamental plants; I suspect the species we have is not one of those chosen.

The next layer belongs to the acacias commonly known as the wattles. In spring the air fills with the aroma of the iconic Australian plant as a succession of species flower over a two month period. First to come out in a blaze of yellow are the silver and Cootamundra wattles, closely followed by the golden wattles, then a final explosion of the large wallowa wattles. There is a sweet characteristic scent of the wattle hanging over the bush and all manner of insects can be seen exploiting the flowers to collect whatever essentials for life they have to offer. Other wattles fit into this cycle, the prickly moses with an impenetrable thorny foliage and many smaller wattles from the silver wattles to some which hug the ground. They all combine to give a hue of golden yellow to the bush which underlies Australia's green and gold sporting identity. The wattle trees and shrubs do not live for long and after a few years die and give the scrubland an unkempt look which to humans may be unattractive but to creatures and other plants provides protection, and as they decompose return nutrients to the soil with interest.

The next spring flower to carpet the bushland is the sweetly aromatic fringe heath myrtle or micromyrtus ciliata bush (my apologies for using the scientific name, I feel the need to demonstrate that I can!). The flowers range from white to bright pink and on to red giving pleasing contrast to the gold of the wattles. When we were looking for another income from the farm we researched the production of wildflowers, taking advice from government horticulturists. Ciliata was a possibility and we took a few to flower shops where the main question was whether the flowers held the inflorescence or dropped them too early for commercial consideration. I don't think the wildflowers liked the idea of being in someone's house in the city as they dropped their bundle within a few days. There went another idea! The combined aroma of wattle and ciliata gives the bush a sweet almost sickly aroma which seems to attract a huge range of flying and walking insects and birds.

Flowering later than the wattle and ciliata is the fringe myrtle which forms a layer of bushes between the two and has abundant starlike flowers in clusters of red, pink, and white. It is the final burst of flowers from our protected area and ends three months of rolling tree and bushy wildflowers which are a delight to behold and a must to protect.

While the main bushes which dominate the plant community are doing their spring thing there are numerous small bushes which fill in the gaps. One is the spiky flame heath with its small, pointed cigar shaped flowers which contrast with the greens of its foliage as if someone had made a permanent flower arrangement. Very hardy, they occupy the lower layers of the understory with great tenacity.

We had a small patch of heath tea bushes which seem to have skipped a few seasons or died out. We will probably find them elsewhere in the bush where they have relocated to a more suitable spot. The sweet smelling wild daphne heath with its white trumpet like flowers intersperses the other flowering bushes adding a contrast to the yellow and pink. Yellow, orange, and purple bushes of the pea family add different hues to the splendidly colourful spring flower arrangement.

Close to ground level amongst all these are the jewels of the crown, the small orchids and sun lilies. In wetter years they carpet the ground in spring, each species rolled out in sequence as if choreographed to give us a grand orchid show. Small in stature but grand in design, the first to emerge are the purple sun orchids followed by the golden and leopard diuris, prosaically named for the black dots on a yellow background. We always wait with excited anticipation for the spider orchids to emerge. First the delicate red spider orchids come up in two or three specific places with their clever design to attract one specific wasp which pollinates the flower. The species is listed as vulnerable and the recorded number of individual plants is small. To add our small populations to the database we had the department of conservation come to geolocate them as a permanent record. We tread very carefully when admiring and photographing them as compacting the soil will affect their viability. The exclusion of livestock is also important and to the irritation of some field naturalists we do not let groups near them. The fact that one specific species of wasp is required to pollinate them shows how finely balanced the ecosystem is and using pesticides in close proximity would probably eliminate the wasp and consequently the precious red spider orchids.

Close to the red spiders is a single small patch of green combed spider orchids. The yellow green inflorescence with red stripes attracts a particular wasp which tries to copulate with the flower and clever design completes the act of pollination. The germination of the seed depends on infection by mycorrhizal fungi which then draws some nutrition from the orchid, another example of finely balanced symbiotic relationships. In the past the tuber has been a food source for indigenous people who either ate the tuber raw or baked.

The succession of orchids continues with the rabbit ears popping up in the barer drier areas and the tiny nodding greenhoods emerging in well sheltered shaded places between the scrub. So small that they are difficult to spot each year making one wonder how they maintain their place in the flora ecosystem. Research suggests that these hooded orchids have evolved over the past 2.5 million years and shown great adaptability to arid climates surviving in niche ecosystems. The orchid season closes with the single erect stem of the onion orchid, which is covered with 8-10 small green flowers, growing from the leaf litter in sheltered areas.

One orchid we would love to find is the ‘flying duck’ orchid. It looks as it is named and has a very amorous way of embracing a visiting pollinating insect in a forceful love-hug covering it with pollen before releasing it to experience the affections of another flying duck which it pollinates.

Throughout the scrub and under the tall trees there are many other small flowers, including the blue chocolate lily and the lowlier but no less important ancient liverworts, moss, lichen and the carnivorous drosera or sundews which catch and digest insects trapped in their sticky inflorescences. The drought tolerant everlastings grow individually with their bright yellow flowers or en masse close to the ground in a carpet of white.

We have encouraged the native trees to self-propagate and where large trees have died they are replaced by saplings with the skeleton of the dead tree left to provide habitat and nesting places for the birds. To mitigate salinity, provide windbreaks and add some colour to areas surrounding the olive grove we planted some deciduous trees such as deltoides and lombardy poplars, elms, and a range of oaks. Apart from the poplars all these have been propagated on the farm from seeds collected from cities. The elms come from Royal Parade in Melbourne and the oaks from Commonwealth Park in Canberra. A bit of a shock to go from those cosseted environments to a rustic Victorian farm, they have nevertheless thrived and add autumnal colour with a backdrop of the silver green olive trees. The wallabies and kangaroos enjoy the shade provided by the expansive canopies and the former enjoy adding acorns to their eclectic diet. The parrots also enjoy acorns and we have found abandoned meals of acorn in the forks of the olive trees.

Soon after our arrival on the farm there was a government push to revegetate denuded farms with native trees. Local councils made seedlings available and we planted some of these in avenues and windbreaks. We soon learned that the survival of trees from other regions was low, something which should have been obvious. Delightfully, some species of eucalypt did survive and we have some magnificent manna, sugar, lemon-scented, spotted, and mahogany gums around the houses. Unfortunately the two manna gums haven't attracted koalas yet. The smell of the lemon-scented gum after a storm is a delight to the senses and the smooth white barked tree we were given when we became Australian citizens is a magnificent specimen of this species. The only other pervasive aroma that is as pleasing as this lemon-scent is the earthy 'petrichor' which fills the nostrils after a rainstorm has provided relief from the baking enervating summer heat. The word, first used by two Australian scientists, is derived from the Greek word for stone and that for the fluid which flows in the veins of the gods in Greek mythology. The mechanism of petrichor is complex involving the release of aerosols and plant oils when raindrops fall on the soil. For those who have endured prolonged droughts there is no other smell which evokes such relief and hope for more rain.

We have introduced over 5,000 olive trees with the varieties coming from Greece, Spain, Italy, France, California, and Israel. There are the main varieties and some which we have planted to test their suitability for our production and management. The core stock we purchased and then to expand we propagated more of the varieties which performed the best. At the same time we collected scions from trees which were growing on farms in the local area and even from trees in the gardens of the local Catholic church. Having no idea of the origin of the latter trees which would give us an idea of the variety, we collectively dubbed the progeny 'the blessed olives'!

In some States olives are regarded as noxious woody weeds as they will grow in harsh environments with little attention and are spread by birds consuming the fruit and leaving the pits or seeds. There are some local 'wild' olives emanating from pre-war horticultural enterprises belonging to Italian migrants. We have collected cuttings from these isolated trees to enjoy seeing what emerges. In the grove there are a lot of 'volunteer' seedlings which grow each year and are a result of cross-fertilisation. Hoping for some serendipitous discovery of a new table olive variety which is a blend we have planted around 600 of these volunteers in hedgerows and look forward to their maturing and providing us with a new variety to be cloned for commercial production.

Apart from the changes the establishment of the olive grove has made to the terrain and diversity of vegetation, the olive fruit has provided a new food source for the birds, wallabies, and kangaroos. It took a while and a couple of droughts for them to cotton on to the new food and they are still discerning in the ripeness and size of the fruit they prefer. The riper and smaller fruit from the frantoio, corregiolo, and leccino varieties are the favourites of the ravens which hang around the grove in flocks once the olives are ripe. They will sit in the tree and break of small branches then hop to the ground to eat the olives. Then the cockatoos and corellas fly in to crack open the olive stones and eat the tasty seed endosperm. The cockatoos, destructive creatures that they are, will break of the branches to help the ravens, a sort of avian quid pro quo. The rosellas will procure their own food and the wood swallows likewise. The galahs don't seem to avail themselves of the olives yet, no doubt their time will come.

Unlike the wallabies which are browsers, the birds haven't taken to eating olive leaves yet. Picking leaves off a tree one by one the macropods leave the tree undressed in a clinical way. Their preference is the kalamata variety leaf and in the fruit line they prefer green olives of the verdale variety. They will strip the part of the tree which they can reach and stand high on their hindquarters to reach and pull down those which are higher. On one occasion the infamous and much loved female wallaby 'Stella' climbed onto a garden bench to reach the olives in an adjacent tree. When caught and admonished she casually turned round still hanging on to the branch and gave us a 'what' look before turning back and continuing her endeavours. You will hear much more about Stella when we recount the relationship which developed with her and other wallabies.

Mowers

Some farmers have a shed full of mechanical equipment such as tractors, chemical sprayers, and mowers. We don't spray so there is no need for equipment to apply pesticides and herbicides. We do need to mow to keep the grass in the grove down and the areas around the houses need to be kept clear of long grass, for this we have the marsupial mowers. There are about 150 of them, they need no fuel or motors as they are kangaroos. Maligned, even despised, and shot by many farmers, we see them as our allies in controlling the growth in the grove and cropping the grass around the houses.

Our local residents are the eastern grey kangaroo, the males of which can grow pretty large, especially when seen in boxing mode stretched to full height and pivoting on their tail to engage with the opponent. The numbers depend on the seasons with breeding occurring all year round being increased or decreased with the availability of food. Their diet is grass supplemented with the odd weed and occasional leaf. Their meagre requirements for water are usually obtained from the grass they eat with very efficient extraction of every molecule in their digestive system leaving very dry fibrous faeces. When the grass is green and moisture levels are high the faeces may be softer as their digestion passes through excess water and in a dry summer when the grass contains little or no moisture they will go to the dams or the irrigation system for a drink. They are animals which are uniquely adapted to the Australian environment.

Normally the females will have a young kangaroo or joey at foot and one in their marsupial pouch. Their elastic pouch can stretch to accommodate an oversize joey with its legs and tail protruding at grotesque angles as the poor mother tries to leg it with this oversize underslung burden. There will be a convulsion in the sac as the youngster manoeuvres its body so the legs disappear and its head emerges to observe the events of the day. The mother cleans her pouch by putting her mouth in the opening and swirling her tongue around the inside, giving the joey a bath at the same time!

The weaning process can take a while with the joey seeking a fill up of nutritious milk as the mother stands patiently while it is at the bar. Over a few weeks the patience wanes and the child learns to fend for itself and become part of the mob. Prior to weaning the joey pops its head out and grazes, learning by association which the best plants are to provide a sustaining diet. Amusing scenes can be observed when the young first venture from the pouch on very unsteady legs and perform unbalanced circuits of the mother, often with an undignified fall as they turn too sharply and the legs and tail get tangled. With a little practice they become adept at hopping and can soon be seen running at great speed doing circuits of increasing distance to show off their prowess. Occasionally they will trip over a branch or irrigation pipe when travelling at speed and tumble in a flurry of dust emerging a little embarrassed and disorientated. The mother observes patiently and waits for the child to come back, often giving it a cuddle with its front paws and the two embrace for a few moments of marsupial love. I have heard the joey communicate with its mother using a clicking sound, a rare and unusual observation.

Going up to the grove to work in the morning we are delayed by the only traffic on the farm as the grove 'mob' of roos return to the haven of the trees and hop over or burrow under the fence in hordes in front of us. There is panicky confusion every day as they find a gate is closed or they leave their escape too late and are threatened by the white monster, our ute 'Sherman', which is approaching relentlessly. Of course we drive slowly enough for all of them to find their way unharmed and disappear in great leaps into the grove. We call this daily traffic congestion 'Hoppers

Crossing' after a suburb of Melbourne where there is daily carnage of commuter traffic.

Along the front of the farm there is a main road which is crossed daily by certain kangaroos which are seduced by the grass being sweeter on the other side. They run the gauntlet of the passing traffic and inevitably get hit from time to time. These collisions are usually fatal with the dead roo lying on the side of the road as a warning to motorists of the dangers of the night, and eventually the decaying carcase provides food for wedgetails and ravens. Sometimes the animals are damaged and we discover them still alive near the road and have to despatch them as a matter of mercy, an unpleasant task at the start of a day. This also occurs when shooters on neighbouring farms go spotlighting at night and invariably wound some of their targets which escape onto our property where we discover them and end their pain. We are always distressed on these occasions and cannot understand the bloodlust of the shooters who seem to think killing is a sport to be enjoyed on weekends. There is no perception that animals suffer and they are destroying a family unit, perhaps they do not think animals have feelings and emotion is solely the domain of humans. If they stopped long enough to observe the result of the violence perpetrated they would see the mother or mate grieving over the body of the deceased.

Recently we saw a young female kangaroo by the sheep sheds which are about 200 metres from the road. She watched us and didn't move as we drove past leading us to conclude she wasn't well. Closer observation revealed she had a broken leg but, given that she had managed to get to the shed from the road where she was probably hit by a car, I decided to leave her to recover and check on her later. When that time came she was gone and a few days later we discovered her hiding in the olive grove and confirmed she had a leg broken at the knee joint. She was eating and getting around adequately, albeit with some difficulty, a shuffle rather than a hop. Lucy was the name given to her and in the ensuing months her leg healed at an odd angle and she continues life with her disability, even putting on some weight. Recently she seems to have teamed up with another damaged kangaroo who has been named Cloppity and they are seen together frequently at the top of the grove.

The mature males spend much of their time challenging the dominant male for mating rites. The boxing bouts are spectacular with the males stretched full length pivoting on their tails while engaging the foe with their front paws and trying to eviscerate the opponent by kicking with their back legs. The battle can go many rounds until one of the combatants retires and retains or cedes supremacy. Often the wounds sustained by this activity and the jumping of fences leaves the male with a permanent injury as is the case with Cloppity who has a crooked back leg which impairs his normal hopping movement. Now he and Lucy shuffle around together keeping each other company.

The collective noun for kangaroos is a 'mob' and this term is generally used to imply that the animals are a homogenous hopping mass. This is simplistic when one closely observes their behaviour on the farm. There are a number of families comprising many individuals and their progeny. Small families may aggregate into larger groups where there will be a tussle for dominance amongst a few large males, the winner constantly under threat of defeat and relegation. Injured or old males live on their own or on the fringes of a group, as they get older they live closer and closer to water to make the daily thirst quenching easier. Occasionally a male will disappear for a while and then reappear many months later, perhaps rested, and ready to resume duties. Such was the case with Lomu, reverently named after the legendary All Black rugby player for his strong muscular build and good looks. Easily identified by a tear in his ear Lomu ruled the grove with his dominant presence for a while until one can only presume that he took time out to recover or perhaps go in search of other interests in neighbouring domains. Behold, a thin and arthritic Lomu arrived at the house to quench his thirst a few weeks ago. A shadow of his former self he let us know he was still alive and living in the bush adjacent to the grove he once ruled. Then he was gone again.

One of the most magnificent sights I have seen was in the dam on a very hot summer's day. Two very well-muscled and tall adonis-like male kangaroos were taking a cooling dip in the water and they stood tall when I appeared in the ute to see the sun glistening on their lean bodies. They stood their ground in the shallows before taking off in unison in a spray of water – memorable.

The mowers around the houses almost completely disregard our presence and continue to crop the grass undisturbed. We categorise them into different classes depending on their proficiency. Numbering up to 20 on any given day there are the general mowers who keep the area cropped, then there is a smaller number of trimmers who graze close in to obstacles such as the bird bath and the garden fences. The most specialised individuals are the edgers who clean up the few blades left by the trimmers and give the lawn a very kempt appearance, even going into the open garage to clean up any grass or weeds daring to grow in there.

It is a great start to the day when I open the bedroom curtains to see the sun rising and a few mowers and their joeys going about their business a few metres away. It puts the world in perspective, the simplicity of their lives devoid of the pernicious labyrinth of human pursuits with which we surround ourselves. Absent are the politicians, bureaucrats, media, banks, and constant pressure to make money, molly-coddling laws and other things which beset our daily lives. Their life is consumed by eating, sleeping, and breeding, simply simple. Not to say there aren't emotional ups and downs with love, grief, envy and sometimes fear of starvation. However, it all seems so much less complicated.

Starvation usually comes with the long hot dry summers and frequent extended droughts. Well adapted to the harsh summers, the kangaroos will slip their young joeys from their pouch if they are short of food, a self-imposed method of population control. They have a unique method of controlling their body temperature on hot days by licking their wrists where the blood vessels are very close to the surface and cooled by an evaporative effect like humans keep water cold in hessian cooler bags. This and other complex mechanisms allow them to tolerate very hot temperatures.

They seem to have few afflictions as we find few bodies of dead kangaroos, apart from the old ones we know, and those shot. There is one rare and nasty disease which affected our mob causing blindness and complete disorientation. Research suggests that the cause is the Wallal virus which is transmitted by midges and causes inflammation and degeneration of the eye and can affect the brain. We had to call the local conservation officer to diagnose the problem and put the few affected kangaroos out of their misery. It was a pitiful sight watching the infected kangaroos hopping aimlessly around the bush crashing into trees and any other obstacle. Thankfully that outbreak in 1994 is the only one experienced on the farm.

Tricky when working around the trees pruning or harvesting are the hip holes which are dug out on the shady side of the trees to accommodate the bony hips of the reclining roo. In shaping these holes any encumbrance such as an irrigation line is moved to facilitate the excavation, which in turn is aligned so that the incumbent can rest the elbows on the tree mounds with the hips comfortable. An added convenience may be an irrigation dripper which conveniently sprays water into the hip hole, the cooling effect adding another dimension to the level of luxury. In this and many other ways the marsupials mitigate the impact of the summer heat on their bodies.

Despite frequent reports of people being attacked by male kangaroos we have not had a major confrontation. We do recall being chased at a distance a long time ago – let's say it appeared to have nefarious intent and we were not hanging around to find out. When doing a bit of tail tickling of an in-season female whom they evidently intend to impregnate given the dangling penis, we give them a respectful berth. They tend to stand tall when we approach in a vehicle or on foot, their way of saying beware, but a clap of the hands usually sends them on their way after the retreating females whom they are notionally protecting.

When we had dogs, German shepherds, they would occasionally disobey commands and set off after a roo with little hope of catching them. However, one did persist and was led to the small house dam where the pursued entered the water and turned to face its pursuer about waist deep in the water. Knowing the fate that was about to be meted out to the dog we frantically and successfully called it off. It is well known that the kangaroos use this defence to entice the dog into the water then calmly use their front paws to push the dog's head under and drown it.

We no longer have dogs or cats deeming the wildlife adequate company and freeing ourselves of the constant need to restrain dogs from the chase and keep cats indoors to stop them killing birds. Unfortunately, a large population of feral cats takes its toll on smaller creatures including birds. Coordinated programmes to trap and euthanise the ferals have some success for a while before they multiply again and become a nuisance.

Stinkers

Stinker is our affectionately adopted name given by Queenslanders to the common swamp wallaby. It refers to their distinctive swampy odour which can best be experienced when on a summer night they congregate under the house and the miasma rises through the floorboards. Their presence is accompanied by the thumping and bumping as they jostle for a good spot or discipline a recalcitrant youngster. Less pejorative names include fern wallaby and black pademelon. They are smaller than kangaroos and run at speed lower to the ground. They cannot pivot on their tails which are long and trail behind them, twitching when excited by a female in season or in a state of agitation.

Being solitary animals they are spread all over the farm and take to the scrub for refuge in bad weather. In the evenings they congregate around the houses where they might find a snack of vegetable cut-offs, rose prunings, olive branches or some other delicacy. They each have a different character and are endearing to the extent that we have had some very special and edifying relationships with some of them.

Our first experience of a wallaby was George who pitched up one day in our vegetable garden and had helped himself to a platter of vegetables in season. To get there he had to clear a considerable corrugated iron fence which we had erected to discourage such incursions from rabbits! We raised the fence. He then became a regular visitor preferring the water from the birdbaths to drinking from the dam – the same way that dogs seem to prefer drinking water from the toilet rather than their dedicated water bowl.

Other wallabies around the farm had their own territory, Culvert lived in the culvert under the road up to the houses and eventually moved up to the olive grove where he spent his retirement. The grove was a safe haven from the bullying of the dominant males by whom he had been ousted from the top of the mating order. No doubt they were quick to remind him of the beatings he had meted out to them in the past. The beatings were severe, I once witnessed a 30 minute bout between the incumbent alpha male, Stallone, and the challenger Ali. They went at it hard with shoulder wrestling and running high kicks to the head and shoulder, a marsupial mixed martial arts of sorts, but more brutal. I watched in awe as it appeared that Ali had won the bout and earned the right to assume ascendancy with his skinned shoulders testament to the intensity of the battle. Stallone must have recovered to challenge the usurper as he was back in command in a few days and Ali was banished to the bush to be seen occasionally when he dared approach the houses. Perhaps this was intelligent use of a strategic withdrawal by Stallone to lull the challenger into into state of complacency.

Stallone joined Culvert, who had an alter ego Poonkie, in the olive grove where they rubbed along eating olives and leaves making occasional forays to the house when the food was scarce. For the latter who really was very old we would carry some treats in the ute and if we hadn't' seen him during the day seek him out to deliver them, a wallaby Deliveroo. We would be quietly working away in the olives pruning or harvesting and we would feel a presence and turn around to find old grey muzzled Poonkie had crept up on his arthritic legs and was standing off expectantly waiting for his daily treat. He was in the grove for a long time until one day I found him with a broken hip unable to move. He must have caught his leg on a root or a broken branch and was in some pain. Sadly I took the decision to end his misery as he could no longer move for water. It is heartbreaking to end the life of a living animal, especially one with which one has built a strong affection and bond. We all know lives will come to the inevitable end but it doesn't deter us from establishing a relationship with them and as with ourselves we hope the end will come swiftly and painlessly after a long life. Sometimes we must end a life to stop suffering, other times we find an animal dead, as happened with Stallone. He became weaker and more arthritic as the years went by spending most of his time in the olive grove with short visits to the houses. That is where I found him dead one day, half under the house on a hot day. I took him up to the top of the olive grove where I buried him next to Poonkie, both graves with big stones over them to stop the scavenging foxes digging them up.

I have no doubt that the medicinal effects of raw green olives and olive leaves prolonged their lives and the oleocanthal anti-inflammatory eased their arthritis as they aged. The curing effects of the olive tree were manifest in the extraordinary story of Stella, a wallaby who came closer to being my friend as any wild animal has. She was a regular visitor to the houses in the evenings and didn't wait for us to give her a snack, she helped herself. Her favourite food was roses, a penchant shared with Stallone, and Ruby, another favourite. Together they caused the fencing off of all the garden flowers, even then any crack was exploited to reach the desired feed. On one occasion the external cat run was opened to let Ruby in to cut the grass, a no exercise convenience. Seeing the opportunity Stallone joined her and picked every leaf off an apricot tree which was growing inside and Stella made up the trio helping with the grass!

We have four olive trees rescued from a potted existence in a Melbourne suburban garden now planted in a small 'grove' with a large garden bench to add some ambience. Of course Stella thought the décor was placed for her use and was seen standing on the bench reaching up to the uppermost branch of the nearest tree picking the succulent green olives for a meal. Her little fingers, as we referred to her front paws, became adept at undoing wire tying fences together, or gates protecting vegetable gardens. It became a game to see whether we could devise a Stella-proof mechanism to keep her out. She could be an ornery beast growling when she felt hard done by and greedily chasing off any competition for food. Despite her very self-oriented character and maverick behaviour she was much loved and her antics the cause of a great deal of amusement. She regularly produced joeys to swell the population, no doubt teaching them all her nefarious tricks.

The trio indulged us with regular 'kisses' which were a sort of spitting motion without the spit. At first we were non-plussed by the action and despite searching the internet and books to this day have not discovered whether it is common place and what it means. As it usually occurs prior to the provision of food we assume that it is some wallaby form of please, or 'I love you if you feed me', or acceptance as part of the tribe. Once I saw a joey blowing kisses to its mother who affectionately put her paws around it, so I assume it is some form of expression of familial affection. Whatever the interpretation, it gives us a warm glow to feel so close to animals which normally see humans as a threat and predator.

One late afternoon driving back from the olive shed I saw an unfamiliar dark lump on the edge of the scrub. Many years in the past I spent an extended stint in the army where we were trained to photographically memorise the terrain so we could detect any change. This lump hadn't been there before and on closer inspection I found it was Stella, emaciated and close to death. I was heartbroken that her end was near and there was little I could do. I gave her some water and some wheat biscuits and cut some rose leaves. She drank a little and nibbled at the biscuits but was too weak to eat much more. I left her fearing she would die in the night or the foxes would find her, thinking that moving her to the protection of the houses would too traumatic and hasten the inevitable.

I didn't sleep well that night and at dawn I went to check on her and she was gone. Presumably she had dragged herself into the bushes to die. I searched in the vicinity for a while and gave up. On the way back to the house I again noticed a dark object nestled against an old tree stump between the two houses – it was Stella. The plucky little thing had dragged herself around 50 metres in the night to one of her favourite spots, what a will to live! Knowing the therapeutic effects of some of the anti-oxidants in olives and that they were soluble, I crushed some olives in water and slowly fed the infusion to her during the day. Olive leaves contain many of the same compounds so she was given those and some rose leaves and flowers. Slowly she started eating them and a little less dehydrated she survived another night, and another, and another. Soon a week passed and we were able to help her to her feet, still unsteady it made it easier for her drink the infused water and she started eating the green olives. Now able to give her a cursory examination it seemed that her teats were inflamed so we diagnosed mastitis. A search of the internet unsurprisingly had many articles on mastitis in humans but not many about wallabies! There was some research which suggested that wallabies didn't get mastitis because they had an anti-mastitis gene which gave them immunity. Later consultation with a vet also cast doubt that it was mastitis. Whatever, we were dealing with a very sick wallaby with swollen teats and herbal (olive) therapy was working a treat. Her appetite improved and she consumed bucketloads of roses, olives, olive leaves, and any other leaf delivered to her bed throughout the day. We would help her to her feet and she started to take a few wobbly steps and then a partial hop and finally after a month she was mobile again.

It was a remarkable recovery which we put down to her stubborn will to live, the therapeutic ant-inflammatory effects of oleocanthal, similar to the drug 'ibuprofen', and the multitude of other anti-oxidants in olives and leaves. During her treatment she completely lost her fear of humans and when fully recovered would accost visitors demanding a treat. Her greatest performance was with a member of the local police force who came to check that Covid quarantine regulations were being abided by after returning from international travel. Standing talking he felt something tugging at his pocket, it was Stella drawing attention to her hunger. The policeman was a little shocked at her familiarity and disrespect for the uniform.

Stella survived and stayed most of the day around my house eating the grass and enjoying treats from the vegetable garden, fruit trees, roses, and the olive grove. She continued to strip the foliage off the lower branches of the trees in the garden and each morning would be waiting for a treat by the back door as I went to work. On a few occasions when I left the door open she came into the laundry in search of food, but finding the tile floor very slippery, retreated to the safety of the grass. If I couldn't see her I would go to the ute and open the door only to find her appearing behind me and grabbing my clothes demanding a treat which she knew I carried a supply of behind the seat.

She lived for many years after her brush with death and became well known around the world through reference to the healing capabilities of olives in conference presentations and social media. She still had solid lumps where her teats were and no longer produced joeys to add to her legacy. She aged and became more arthritic, less able to cope with the hot summers and eventually could no longer stand as her hips gave way. For a few weeks I moved her around the garden from shady patch to shady patch, fed her roses and showered her with water to cool her down. One morning I found her shaking with fear and desperately trying to get away from me, I concluded that the fox had visited in the night and she was petrified. With a very heavy heart I ended her misery and she is buried under her favourite olive trees in the garden with a large pot containing a red 'Friendship' rose as a headstone. I still talk to her when I water the rose in the summer. She taught us to observe the farm animals and befriend them as equals with the same strengths and frailties, emotions, and characteristics as humans. She was a beacon in our lives.

We are reminded of her when we experience a trait from one the wallabies which we suspect was inherited from her. As I write there is a thump and rattle in the area of the vegetable garden which abuts the house. The green luscious food attracts much interest from the wallabies and is therefore protected by a reasonably high fence. Until now the fence has effectively kept the stinkers at bay. We have a feisty young male wallaby we call Sammy J, named after an equally feisty Australian comedian. Sammy J looks very like Stella so we assume direct lineage, probably assisted by Stallone. The rattle and thump was Sammy J jumping the fence and when I looked through the window he was devouring a tomato plant. Who could blame him when one looks out at the parched grassless paddock outside the garden. He took off back over the fence when I admonished him, only to jump back and carry on within minutes. Having seen where he breached the fence, I was able to raise it and frustrate him once again! He has carried out a full inspection of the fortifications and no doubt is already scheming on how to breach them again. Ruby was a character. With a greying roman nose she would stand sweetly blowing kisses encouraging us to give her snacks. She was more genteel than Stella and had far better manners, and with Stallone they formed the trio which was a seminal generation in our relationships with wallabies. Ruby died, we think from a stroke, and predeceased Stella. She is buried in the grove next to Stella with her favourite taste, a Papa Meilland rose, in a large pot as her headstone,

The number of wallabies has reached almost 20 from the original George. They all have names and individual characters and we watch the battles for ascendancy among the males. The females are less caring of their offspring than kangaroos who will look after the joeys for many months once they have been evicted from the pouch. The wallabies turf the joeys when they are much smaller and do not nurture them for long after that. The result is that there are some small and nervous youngsters, who having observed from the pouch that there is food at the houses, come in in the evening and patrol around the periphery hoping for a snack to be lobbed in their direction. The smart ones learn that feeding time at the different houses varies and if they get their timing right they can enjoy two snack sessions in one day.

Our relationship with the wallabies is an indulgence and I can hear a cohort of naturalists shouting that we should not feed them. Firstly – why not, secondly, they have taught us much about their behaviour and encouraged us to observe that of others so have served as a link to the wild world. The feeding is very much on their terms, when their favoured diet such as wallaby grass and the leafy shrubs are available in the bush, they don't come in for a supplementary snack, so we have no fear of them becoming dependant on us for food. It is a peaceful coexistence where we give them a treat and in return they give us immeasurable pleasure and companionship.

Blowies, Mossies and Mice

I can find very little that is endearing about flies. I know they play an important role in the ecosystem but, dear Creator, isn't there something that can take their place and be less intrusive and irritating? Even if Noah refused them passage on his Ark they would cadge a lift on the animals, breeding in multitudes ready to populate whatever landfall the vessel made.

The blowflies are the most invasive with their acute senses detecting decay or meat cooking on a stove from great distances. They are able to find any crack or slight opening in a house and creep in to buzz irritatingly until felled by a can dispensed pyrethroid. In the open they will lay their eggs in the wet wool of a sheep or the decaying flesh of a carcase to hatch into maggots and more flies in a very short time. The farmers who agist their sheep on our property are very diligent about checking their sheep for fly strike during the hot wet periods and quickly clean the wool up. There are some farmers who do not inspect their sheep so frequently and the fly blown sheep lose their wool and eventually their lives. The kangaroos and wallabies do not seem to be affected unless they are infirm and cannot reach infested parts of their body. The blow flies bide their time until they can infest the carcase and multiply.

I accept that their role is to help clean the landscape of deceased animals, what is left after the foxes and raptors have had a meal or two, but do they have to be so active in their search for a victim!

Smaller and more irritating for those working in the groves are the bushies or bush flies. They seem to have little purpose other than to carry some fairly nasty bacteria like Salmonella. We have spates of them in summer, so many that we sometimes wear flynets to cover our faces and stop them from pestering our eyes in search of moisture. They breed in dung, mainly that of cattle and I have not seen any maggots in kangaroo, wallaby, sheep, rabbit, or hare faeces – they are all too dry. Their number depends on the activities of dung beetles, in a season when the rains and dung beetle breeding coincide there are fewer flies, if the rain is early the flies abound until the beetles roll up their balls of dung and fill them with eggs. So we can blame the outbreaks on the local cattle, of which thankfully there are few, horses, and perhaps the dogs of which there are not many. We have resorted to using many fly repellents which are on the shelves at great expense and mostly ineffective. The best seems to be a creamy concoction of botanicals including tea tree oil and some pyrethroids, and a strong wind helps.

Until this year we thought the biggest fly we see is the horsefly which moves at speed looking for some exposed flesh to bite and suck the blood they need for egg-laying. Fortunately, they are rare and we have clothing covering our bodies except arms and faces so do not provide a target for their painful bites. There are few horses around to make the area attractive to them so invasions are infrequent.

This year we were buzzed by a new member of the fly family, the appropriately named robber or assassin fly. With a profile somewhere between a house fly and a dragon fly, these flies cut through the air like a Blackhawk helicopter and make almost as much noise. They are voracious predators with our first sight of one landing on the ute embracing a locust on which it was feeding. Apparently, this large menacing fly can inflict a very painful bite. Mercifully we have been spared this experience. Seeming to frequent the treed gullies with short feeding forays into the grove they are no longer around – perhaps having laid their eggs to emerge in spring.

Hover flies are beneficial insects which often look like bees or wasps. They appear in large numbers in the olive grove and bush in spring and summer where they pollinate flowers and predate on aphids and scale insects. They do not attack, sting, or pester humans and are all together a nice fly – according to the books there are about 160 species of them in Australia. I have no idea how many species we have on the farm, I would suspect a few.

Back to the biting flying insects, we have the sandflies and midges around the dams, and mosquitos everywhere. Carrying precursors of some serious afflictions the mosquitos do attack us and come in various shapes, sizes, and colours. They all have that high pitched stuka-like whine as they dive in on your body to suck blood at night. The usual evidence of the involuntary blood donation is an itchy inflamed area of skin where the mossie has injected anticoagulant to keep the blood flowing during its feast. However, they can carry some quite nasty diseases which they leave in return for your gift of blood – and I caught one of them. Ross River Fever is imparted by a virus and it is a painful uncomfortable affliction. First the body breaks out in spots, or a rash as the medicos call it, then every joint in the body feels arthritic, especially those in the arms and legs. Moving is painful if you have the energy to move as tiredness is another symptom. It knocked me out for about three months with the doctors offering little in the way of cure or pain relief. The best remedy came from a friend who said that drinking a lot off gin and tonic was as good a cure as any, I took his advice.

Thankfully we do not at present have the problem of olive fruit flies which beset the groves in Europe. The adults lay eggs in the olive fruit and then hatch into larvae which eat the olive fruit rendering it useless for table fruit and severely affecting the quality of olive oil which is deemed to have a grubby defect. We hope the increase in the host plants in our area does not encourage the flies to migrate from the north where they cause considerable commercial damage to fruit production.

I have put the rodents in the same chapter as the flies as they are more irritants than lovely endearing creatures. I am sure Sir David Attenborough would describe a useful purpose for them in the natural world, may be even worthy of a documentary, but apart from the antechinus or pouched mouse we call Roly they do not attract our affection.

I have described the mouse plagues in an earlier chapter and rats appear from time to time. Our fear is that they will eat the plastic off the electric wiring in the house and up will go the wooden houses in a puff of smoke. We tend to eliminate them.

I experienced the impact of a rat eating the coating of electric wires in my beloved red Alfa Romeo car. Driving in winter we tended to park the cars facing into the garage which meant that the engine was nice and warm after a run and a cosy nesting place for a rat. Doing a routine check under the bonnet one day I discovered a rat's nest on the engine and the rodent had pretty well stripped all the plastic off the loom of wires which controlled the electrics of the vehicle. How there wasn't a fire started with the grass and other litter collected to make a comfortable nest I do not know. I saw dollar signs ringing up as I imagined replacing the loom at a garage would cost a bit, more than I wanted to outlay at that time. I am not mechanically minded and do not enjoy working in small spaces with my hands – but I decided to try to repair the damage by reconnecting everything with new pristine pieces of wire. Luckily the Italian engineers who designed this sublime car had colour coded the wires so I could work out which to connect to which. Seven hours later all the wires were reconnected and with the help of the owner's manual I rebooted the computer and hey presto, the car started and all the lights and electrics worked, except the cigarette lighter which was redundant anyway. Well pleased with myself, I confessed to the Alfa Romeo service centre that the loom might need replacing – he remarked that he had noticed the 'Heath Robinson' fix and saw no need to replace it. Job done and it has not been replaced to this day!

I don't know whether Roly was the perpetrator, he seems too cute with his shrew like features and fur to commit such an egregious act. The antechinus is a marsupial which hops around rather than running like the house rat. We are not sure whether he is a he or she, male antechinus spend their short life looking to mate and then die soon after the act. The females live a little longer and may reproduce three or more times. So we don't know if Roly is a female waiting for a mate or a male forlornly looking for a female. Apparently the act of mating is quite frenetic, hence the demise of the male, and a female may mate several times in one season thereby despatching a number of males. Perhaps such an act would qualify for a misandrist accolade. We haven't seen any little Rolys so the gender mystery continues. I do know that the animal has a penchant for shoe polish, especially when delivered in squeegee dispensers. A shelf of them was completely demolished with the plastic being chewed through and the polish devoured. I did notice the faeces had a polished look and Roly didn't die because it took him a few days and then he moved on to devour other plastic containers.

Invaders

Rabbits, hares, and foxes are recent inhabitants of the continent. Introduced for various reasons they have had a profound impact on the environment. The hares, reputedly introduced in Victoria in 1862, are the least invasive being fiercely territorial and solitary they do not breed as prolifically as rabbits. They are surface dwellers and do not burrow causing little disruption of the ground. They do cause damage by gnawing the bark of trees resulting in ringbarking and death of young trees and they will gnaw through the stems of small shrubs. Our residents chewed through a few young olive trees causing us to surround the trunks with tree guards. Apart from this disruptive behaviour they have little impact on the farm and provide prey for the wedgetail eagles, foxes, and larger feral cats.

Rabbits are a nuisance and can denude an area of vegetation in a short time. Their burrows are ankle breakers and can spread over a considerable area. Foxes and cats will stalk them and eliminate a few but cannot keep up with the prolific procreation of the animals which came out with the First Fleet in 1788 as a food source. Later they were released for hunting, a practice which continues today with ferrets and guns. Rabbit meat appears on some menus being sourced from commercial rabbit production which is a controversial form of farming given the potential for escape and exacerbation of the existing problem.

Draconian methods have been used to control the populations including the use of poisoned carrots spread over the ground. This unpopular method has an impact on predators such as wedgetail eagles leading to public disquiet and calls for the practice to end. Viruses have been introduced, the first being the myxoma virus which causes the disease myxomatosis spread by mosquitos and other flying insects. Initially mortality rates were high, reducing as the rabbits developed immunity and populations started to grow again. Another virus, calicivirus, was released and stemmed the population explosion for a while until again immunity provided protection.

We refuse to use poison on the farm and do not shoot rabbits. The population is confined to the bushy areas and the odd warren appears under the sheds. The predators and natural attrition seem to control the numbers to levels which are tolerable.

Fox cubs are cute and at times we see three of them around the houses with one or other of their parents. The introduced animal is not popular being seen as a major predator of native wildlife and killer of lambs. Farmers pride themselves in shooting them and hanging them on fences close to main roads in a display of macho human dominance to impress their neighbours. Regularly the parks and wildlife department put a bounty on them recoverable on the presentation of scalps or tails. We accept that they are destructive and have mixed feelings about the pair we have on the farm which are attractive animals. There is no doubt they kill young kangaroos and wallabies, but they also help to keep the rabbit population under control. Admittedly we do find the odd lamb leg abandoned around the farm, however, it is difficult to determine whether the lamb has been killed or was stillborn.

Grass and Weeds

Grass seems a mundane subject compared to the dynamics of animals which graze it. A map of Victoria identifying areas of remnant native grasses shows how depleted the stocks are with farmers planting pasture to sustain the livestock and legumes to replenish the soil between cropping cycles. We are guilty of introducing new species such as phalaris, rye grass, and cocksfoot to bind the soil which can be bared by grazing and prone to erosion. Clover has been sown in the olive grove to improve the soil through nitrogen-fixing bacteria contained in nodules on the plant roots.

In the year after establishing the olive grove, which involved considerable earthworks and relocation of topsoil to make ridges for the trees, we had heavy rainfall which caused substantial gully erosion. To stabilise these areas we planted a cocktail of grasses recommended for rapid growth to bind the soil. These added to the range of grasses which can be found on the farm. However, there is the large area of scrub where native grasses are present, including wallaby grass and many we have not identified. Fortunately the introduced species have not encroached on this area yet.

With the relatively poor soil structure the grasses in the pasture areas are a mixture of the introduced grasses with spear grass, windmill grass, barley grass, and many others which are of low nutritional value for livestock. Management by the neighbours who agist their sheep aims to keep the pastures grazed without denuding the paddocks and exposing them to erosion. During droughts this is difficult with supplementary feeding necessary and the ground can look depressingly bare.

Sedges, onion grass and nut grass are found in wetter areas where the soil suffers from waterlogging. The cockatoos, corellas and galahs will spend hours digging up the onion grass cultivating large areas in a day.

Other invasive weeds are erodium or corkscrew, the seeds of which look like corkscrews and are very effective in imbedding themselves in socks. Pattersons curse or salvation jane with its attractive purple flowers has made a few forays and cape weed carpets the olive grove in spring with its beautiful yellow daisy inflorescences. The very prickly and sticky bathurst burr appears from time to time along with the equally prickly scotch thistle, one of many varieties of thistle which tend to cover the ground around trees where sheep have taken to the shade and fertilised the soil.

Every year a few days are spent digging out the horehound plants which have re-established themselves. White horehound has reputed medicinal uses, especially against respiratory ailments, and has found its way into supplements and even candies used for throat infections in the United States. It is given no such respectful status on the farm where it spreads invasively and the seeds are picked up by the merino sheep and contaminate the wool. We have a small patch of cape tulip which we also dig out annually. It produces a pretty yellow or salmon-pink flower and only has one leaf and is toxic to livestock, containing glycosides which affect the heart. This crafty escaped garden plant has two ways of spreading, via corms or bulbs and seeds. When you pull it out you think you have the whole plant with the bulb, but what you see is the seed capsule and the corm stays in the ground to await the next year! The seeds can remain in the ground for seven years before germinating contributing to the saying for all weeds, 'one year's seed, seven year's weed'.

Many of the weeds have natural predators and their flowers help sustain the life of insects, an example is the flower of the cape weed which is a regular destination for bees. A particularly vigorous plant is the malva or mallow with its prolific purple flowers and seeds called 'cheeses'; it grows into a mass of tall woody weeds. The plant extract has anti-inflammatory and anti-oxidant properties which are used for natural treatments by some cultures, the leaves are also used as a healthy vegetable collected by ardent foragers. For many years we would whippersnip the young plants to stop the weed spreading, then we noticed that the plants we missed and grew to a mature size were attacked by some natural predator which perforated the leaves and the plants eventually died. It made sense to give these predators enough plants to multiply and feed on so we stopped removing the malva mechanically. Ergo, in three years the malva plants were no longer a problem.

Wild oats are rife in the pasture and cause our neighbours to grumble about the seed infesting their crops over the road. Not only does it compete very effectively with the sown cereals but its grain also contaminates the harvested crop sometimes resulting in the product being downgraded and a lower price paid. As we do not use chemical controls we rely on grazing management to keep the growth down and mollify our neighbours.

The next anecdote reinforces the dynamism of all the species of plants and animals which live on the farm. The grass and weeds provide food for all manner of insects, spiders, birds, and the macropods, as well as providing habitat for them and nesting sites for the ground-nesting small birds. When we were forming the grove the contract grader driver could not understand the hat which was put on the ground where he was heading, forcing him to change direction. We had found a small nest with four eggs in the grass and the hat served to protect them. We became adept at noticing the birds living in the grass running and taking off as we approached, tracking back to where they started running we would find a nest with eggs or chicks and mark it with a stake to ensure they were not crushed by the vehicles or a human boot. The number of these nests varies with the amount of groundcover, with droughts depleting this habitat and affecting the population of small ground dwelling birds.

Everything Else

Not everything fits neatly into plants, animals and the soil applied divisions when describing our farm. There are some overriding activities which affect everything, some instigated by us and some which are out of our control such as fires and storms, floods, and droughts.

Fires are a regular threat and the emergency alert system works overtime in summer. There are local alerts for grass fires started by harvesters running the headers so low that they cause sparks which ignite the tinder dry stubble, or overheating bearings which do the same. One local farmer is a recidivist starting numerous small fires close to our farm almost every year. The imperative to get the grain off on hot dry days is understandable, whereas taking the risk of burning the crop and stranded equipment is not. Luckily these fires are soon put out by the volunteers of the Country Fire Authority which have fire trucks and crews dotted around the locality. Another way farmers spark conflagrations is by stacking hay which has not completely dried out and has a high moisture content. Bacterial action generates heat which results in spontaneous combustion destroying the haystack and shed and threatening the surrounding farms. Difficulty in getting water to the seat of the fire makes these fires difficult to put out and they can smoulder for days.

During the spring and summer one of the common causes of fires is lightening during dry storms when the moisture evaporates before reaching the ground, a phenomenon known as virga. The only occasion when fire threatened the farm occurred a few days after a dry storm. It was a hot day with the very hot and dry north wind gusting across the farm when I was relaxing in the house with the air conditioning on. Suddenly I heard the crackle of radios and the chop-chop of a helicopter with a background of moving vehicles. Then I smelled the dreaded smoke and went out the back door to see what was happening to be greeted by my neighbours saying a fire is approaching and I better get out now.

We have a plan for these occasions which includes switching on an irrigation system with pre-positioned sprays to wet the house and lawn, grabbing a couple of pre-packed bags of special things and two cats in cages, taking vehicles out of the garages to be parked in the open, and then getting the hell out of there, fire permitting. I did all these things and driving out saw the fire had jumped the main road and was burning rapidly towards the houses and the bush behind them. Fortunately for us three things happened at once, the wind dropped momentarily, three new fire trucks arrived with tanks full of water to supplement the depleted supplies of the trucks which had initially engaged the fire, and the helicopter arrived with a bucketload of water which it dumped on the leading edge of the fire. Within minutes the fire was under control and council graders cleared a fire break ahead of the front while the fire crews blacked out the burning debris on the periphery. The air was filled with acrid smoke and would be so for at least a week. That night looking out of the front window it looked like I imagined a defender of a castle would see as the fires of those laying siege passed the night keeping warm.

The immediate danger may have been over but there was still work to be done as dead logs and underground dead roots smouldered for weeks and on two occasions the wind got up and the fire started to break out along the gulleys. This was facilitated by the eucalyptus trees dropping all their leaves as the first phase of their recovery from the fire. The wind blew the leaves into the gulleys where roots were burning underground and where the root came to the surface it set the leaves alight and like a fuse the fire sped towards the grassland. For almost a month the neighbour and I went out during the day and night with trailers of water to put out stumps burning deep in the ground.

Later we heard that the fire had been started by a lightning strike on a dead she-oak tree about four kilometres away. The tree had smouldered for four days and broke out when the hot north wind increased speed and blew the fire towards us. We were fortunate as there was little burned other than a bit of fencing, some pasture, and some old trees. The fire came to within 200 metres of the shed and 300 metres of the houses, the olive grove was a little further away. This is an example of an event which can have a devastating effect on the balanced ecosystem we have encouraged.

One evening I was sitting having an early supper when I looked up and saw flames on the main road about a kilometre away. I called the fire brigade and the response was quick with the fire extinguished before it could spread. We concluded from the tyre marks on the road that an arsonist had lit the fire and sped away. Over a three month period we had our fences regularly cut along the main road during the night. The perpetrator would travel some distance to cut fences along a few main roads to satisfy some vandalistic urge. Repairing the breaks took time and resources so when he and his accomplice were eventually tracked down and prosecuted through some good policework we were relieved.

Dangerous fire seasons and droughts usually coincide and are common occurrences to which Australian flora and fauna are well adapted and resilient. Following droughts there are usually wet seasons which replenish the water reserves and give the opportunity for the flora, then the fauna to recover and proliferate. We welcome rain even if delivered by super-cells which resemble mini tornadoes and can uproot huge eucalypts in their path. Floods don't really affect the farm as the houses are on high ground and the country is undulating. A creek which runs through the farm, Foley's Creek, does flood and prevent access to parts of the farm when it is in spate.

Windy storms break large branches from trees and inconvenience the inhabitants, including birds which nest in the hollows. We have rescued very wet and cold possums disturbed from winter hibernation and after a warm night in the house in a cat cage released them back to the same tree. They ungratefully take off at speed up the tree without a backward look to acknowledge their rescuers. Young magpies are also blown out of their rudimentary nests and some have become household pets before being released back to join the army of magpies which dominate this earthly space.

The farm is not a place where arachnophobes will feel comfortable. I have left the spiders till last to avoid losing readers who may have unpleasant dreams after reading about them. The most magnificent are the golden orb weaving spiders which can reach several inches in diameter and despite their formidable size are harmless to humans. They spin large webs with very strong golden thread, in wet years in the olive grove there seems to be a web in every gap along the row of trees. We prune and harvest carefully so as not to disrupt their webs unless absolutely necessary.

The thread from the golden orb has been used to weave a unique shawl by Simon Peers, a British art historian and Nicholas Godley, an American. The pair were following the endeavours of a missionary in Madagascar in the 19th century who started a business of making textiles from spider 'silk'.

While the golden orbs attract the greatest attention because of their size, the redback spiders receive the greatest respect because of their venomous bites. Having been bitten by one on the elbow I can attest to the pain caused and the lengthy time taken for the swelling to subside. Apparently the redback has a bizarre mating ritual whereby the much smaller male is devoured by the female, on some occasions while the mating is in progress. Even more bizarre is that the male places itself on the female mouthparts inviting cannibalism. There are various explanations for this behaviour which sets the redback apart from 'normal' animal behaviour. The spiders are mainly nocturnal and prefer protected spaces rather than the open. Thus we are cautious when moving items in the sheds, often seeing redbacks with their eggs.

While working the olive trees we see many different species, from medium sized well camouflaged bark-like spiders to tiny green ones which are the favoured food of the wrens. Once a year we have the phenomenon of long wispy spider threads flying in the wind, each carrying a minute spider to a new habitat. With closer inspection of the trees to detect olive lace bug eggs we have discovered species for the first time, an example being an attractive green oval spider with red 'buttons' along its back reminiscent of a waistcoat. This new discovery was devouring some lace bug eggs and nymphs so it is very welcome!

In the houses, especially in winter when they seek warmth, we regularly have huntsman spiders decorating the walls. We often relocate them using nets and deposit them outside. Our guests generally do not like the idea of sharing a room with large spiders so we tend to plug any holes through which spiders might enter.

Feared and sometimes hated by humans, spiders are as important to the ecosystem as we are. They eat flies which in my book makes them an indispensable part of the ecosystem – unlike humans!

Where to Next?

I started this story with an accolade for Terry Simpson and the legacy he has left with the genesis of Landcare. He started something because commercially it was sensible and he thought it was achievable. After recounting the development of our property over the last 40 years I perceive there are two fundamental differences in our approach. Terry worked with government departments and courted the support of politicians to achieve his vision. He succeeded; however I believe achieving our vision on a large, even national, scale will be very difficult given the commercial considerations of an approach which challenges the norms of modern day agriculture – high tech with large yields needing powerful machinery, substantial inputs of fertiliser, herbicides, and pesticides. This is a high risk approach which contrasts with our low risk ethos optimising income from manageable on farm production supplemented with vertical integration into the services related to our enterprise.

It will be difficult to change the attitude of the majority of farmers and, for that matter, city dwellers to accept that pests aren't pests and weeds aren't weeds, both can be harnessed to replace some of the inputs or fashioned into saleable products. In a society where the political decision-makers are isolated from the day to day activities of their electorates and largely influenced by lobbyists, environmental considerations come in the form of sweeping statements on pending global climatic disasters. Lost is the understanding that changes on a global scale start with changes at the local level which spread like the ripples in a pond.

We do not set out to change the world, our relationship with the farm ecosystem is more of an indulgence giving us daily pleasure from the antics and seasonal cycles we witness, promulgation in a political sense is not a pursuit which we contemplate. Visitors to the farm from around the world, and even some locals, feel the peace of the environment and marvel at the simplicity. It is a far cry from their bustling lives driven by deadlines and competition, it is back to the basics of nature with as little human interference as possible. There is a price we pay, summed up by a local real estate agent who, when discussing the potential for sale of the farm, was only interested in the arable area and the ability to clear the bush to turn it into arable ground. In his view the ecology has no value, a reflection of the attitude of his potential purchasers. It is a sad day when we have to contemplate selling the farm to someone who will eliminate the animals and destroy the wildflowers to replace them with a monoculture of cereals and sheep.

We could place a covenant over the property which rigidly controls the activities that can be undertaken. Having looked at this it is too prescriptive and preserves the farm as an ecological entity, rather than a commercial activity where a reasonable living is earned in concert with the natural life around us. We could join the exhortations for consumers to pay a higher price for produce from farmers who sacrifice the higher yields of intensive cropping to meet a self-imposed environmental consciousness. We have tried this through certifying our products as organic to find that the premium prices are not sufficient to cover the burgeoning cost of compliance and lower yields. Rather than push at the retail end our approach has been to moderate our expectations at the production end and reduce input costs to a bare minimum. We live well, travel overseas annually, and have a business which is interesting and satisfying. We can only do this with the development of off-farm activities which are founded on our intimate knowledge of the olive grove and the products it produces. This is far more sustainable than depending on institutional handouts which ebb and flow with the current political commitment to ecosystems.

There will always be change which will require farmers to keep informed and able to adjust. There is no such luxury for the animals and plants which have to modify their existence to survive when humans change something to benefit a small vocal segment of the population, keeping scientists and politicians happy so the money continues to flow. There seems little regard for the impact of these changes beyond the supermarket shelf or doctor's surgery. An example is the response of the eminent Australian research organisation CSIRO to the gluten intolerance of a cohort of the population – arguably the more well-heeled. At a seminar in Melbourne where CSIRO leadership was presenting a summary of their groundbreaking research to the food industry, they announced that they were well on the way to breeding gluten free wheat to overcome gluten intolerance in humans. Data on the incidence of the affliction varies between 1.5% and 5% of the global population. To accommodate prevention we are looking to change 100% of the wheat to be gluten free. This was deja-vu for me and took me back to the glyphosate scenario in Singapore. The presenter from CSIRO was surprised when I asked, 'what about the effect on the parrots and other seed-eating birds'? More than surprised she admitted they hadn't considered that and gave me the usual 'I will get back to you' brush-off. Of course she never did and the question remains unanswered. Perhaps we should celebrate the evolution of gluten intolerant parrots to parallel human evolution. Stretching the imagination a bit we could have 'glyphosate ready gluten free wheat' – whoopee!

Recently, big pharma has publicised a pill which can be taken with a meal to neutralise gluten intolerance. Does this render the gluten free wheat redundant? Perhaps we need another pill for the cereal eating birds with stomach complaints!

To be fair there was a presentation which seemed to be more in keeping with an ethos of sustainability and feeding the world – apart from the gluten intolerant. A process has been developed to collect food waste, particularly vegetables, and dehydrate it to a powder which can then be reconstituted – just add water. Light and easily distributed life sustaining food, which would otherwise end up in organic waste dumps, can feed the impoverished, now that is worth investing in!

While we do not promulgate, there just isn't enough time, we do advocate that anyone looking for a life closer to the natural world with the excitement of daily entertainment from wild creatures and plants follows our lead. There is some toil involved, however, it is in the open air where the odour of fumes from a tractor can be smelled from two kilometres away, compared with a city street corner where the fumes are the atmosphere diluted by fresh air. One could probably enjoy the aroma of freshly brewed coffee from the same distance – now there is a thought!

Perhaps selling the farm to like-minded people is the only way we can preserve the equilibrium of this small patch of earth. We may have to accept a discounted price, or paradoxically, someone may be prepared to pay a premium for this unique set-up. Then we can pass on the names of all the mowers and stinkers, show the location of the banksias and spider orchids, and give insights on how to manage the olives to provide the cash flow.

There is no doubt that whoever takes over the property, new owners, or family, will be beguiled by the charm of the daily lives of the creatures and the seasonal rhythm of their surroundings.

Made in United States
Troutdale, OR
11/03/2026

56844492R00069